From **HAND AXE** to **ZEUS** to **QANON**

Our Ancient Brain:

Why We
Think
the Way
We Do

James S. Bucko

"Belief, superstition, conspiracies... How did they get in our brain? This book lucidly explains how it happened." — G.H. Goldsmith, MD

H
HALLARD
PRESS

Cover Design, Typography & Production by Hallard Press LLC/John W Prince

Published by Hallard Press LLC.
www.HallardPress.com   Info@HallardPress.com   352-234-6099
Bulk copies of this book can be ordered at Info@HallardPress.com

Publisher's Cataloging-in-Publication data

Names: Bucko, James S., author.
Title: Our ancient brain: why we think the way we do / James S. Bucko.
Description: The Villages, FL: Hallard Press, 2022.
Identifiers: LCCN: 2022905044 | ISBN: 978-1-951188-49-8 (paperback) | 978-1-951188-51-1 (ebook)
Subjects: LCSH Brain--Evolution. | Neurosciences. | Neuropsychology. | Thought and thinking. | Cognition. | BISAC SCIENCE / Life Sciences / Evolution | SCIENCE / Life Sciences / Human Anatomy & Physiology | SCIENCE / Life Sciences / Neuroscience | MEDICAL / Neuroscience | MEDICAL / Neurology
Classification: LCC QP376 .B83 2022 | DDC 612.8/2--dc23

Printed in the United States of America   1 2

ISBN: 978-1-951188-49-8 (Paperback)
          978-1-951188-51-1 (Ebook)

HALLARD
PRESS

Kudos to the author for his deep, unbiased research and the way he delivered these complex topics.

There are many books on evolution, on the brain, and the psychology literature is inundated with publications.

Few authors have, however, tried to dissect the concepts of how evolutionary biology affects our acceptance of religious beliefs, and influences our bias and misconceptions.

In his book *From Hand Axe to Zeus to QAnon*, Jim Bucko displays his talents and knowledge on the topics of anthropology, neurobiology, evolutionary biology, and psychology in putting together this deep, but readable book on the topic.

The author delivers an animated story on the development of belief, misconceptions, and superstitions. He takes us through the process of how our brain's "rational thinking" is guided by impulses and emotions.

Rarely have I read a book that upon finishing it, could not wait for the time to read it again. It will be a great addition to the library of anybody in search for knowledge. Kudos to the author for his deep, unbiased research and the way he delivered these complex topics.

Jose A Gaudier MD
—Neurologist and Evolutionary Biologist

Introduction

This book is the result of a chance archaeological find in Israel in June 2016 while I was participating in the Jezreel Valley Research Project (JVRP) with the Albright Research Institute in Jerusalem, as I did for several summers.[1] The goal of the project was to survey the entire valley for evidence of human activity. We documented pottery sherds, stone tools, bedrock carvings, stone walls, and other artifacts. This was a relatively new type of archaeological project for me as my previous projects involved excavations at single sites.

My find was a Paleolithic hand axe from the Acheulean[2] tradition of tools that were in use across Africa, Asia, and Europe from around 1.7 million to 100,000 ya (years ago).[3] It was a bit of a spiritual experience for me to hold something made by one of our ancestors perhaps a million years ago. Prior to that, the earliest artifact that I had ever held was only 4,000 years old. The hand axe was properly identified, registered, and boxed and is sitting in a basement somewhere in Jerusalem getting older (Figure 1).

The event piqued my interest in stone tools, tool-making, and how we humans got so good at it compared with other anthropoids in our evolutionary bush. I also began to read journal articles on experiments involving fMRI (functional magnetic resonance imaging) of brain function when a person manipulates an object or imagines doing so.'

Figure 1. Stone hand axe found by the author.
Image credit: photo by the author.

These articles launched me on a journey of discovery regarding the evolution of brain development, the differences between primate brains and human brains, and the impact of these differences on behavior. As Dietrich Stout observed in a recent article, "Stone tools provide some of the most abundant, continuous, and high-resolution evidence of behavioral change over human evolution."[5] Our tool-making capability seems to have preceded and then paralleled other developments in language, art, ritual behavior, and so on. I began to wonder how our species evolved from chippers of stone tools into architects of artificial intelligence. Israel has a rich history as a transitory area for ancient hominins and is also located in the region where early civilizations and the three major monotheistic religions developed. Further questions regarding the connections among tool-making, the development of cultures, and the beginnings of religion occurred to me:

• Why are *Homo sapiens* the only animals (that we know of) able to make sophisticated tools, enact complex rituals, invent religions, and believe in conspiracy theories?

• How did early rituals and beliefs in gods develop?

• Why do so many modern *Homo sapiens* continue to believe in conspiracy theories, religions, and magic?

• Are human beings born with a propensity for religious belief?

• If so, why, and how did it evolve?

Consulting research in many disciplines—including anthropology, philosophy, theology, psychology, and religious studies—I was aware of the various theories that have been advanced regarding the evolution of tool-making and religion. However, as is the case in much academic work, the individual rails tend to be of a narrow gauge. It occurred to me that each of these disciplines has a contribution to make in answering the questions that I was posing. However, the developments in neurology and, in particular, empirical studies using fMRI to identify the areas of the brain involved in making tools, imagining situations, and engaging in religious activities such as praying have provided an enormous leap forward in understanding neurological and behavioral connections. Humans acquired these capabilities through an evolutionary process, and the fossil record offers some clues as to when and where this process occurred. Insights into these issues are also to be gained from the study of some of our close genetic relatives (i.e., the great apes), contemporary hunter-gatherers, and the development of the brain from birth through childhood.

Comparisons of human DNA with that of our closest relatives, chimpanzees and bonobos, indicate an overlap of around 98%. However, this amount of overlap allows for enormous differences in behaviors and capabilities, with the making of sophisticated tools, development of verbal speech, building of complex structures, creation of art, and display of complex ritual behavior being limited to humans. If other species have religions or belief systems, humans have not documented them.

My initial inquiry revealed that the enhanced cognition or consciousness that facilitates self-awareness is a prerequisite for tool-making. Some other species, such as chimpanzees and crows, make rudimentary tools[6] that they employ to acquire food or, on rare occasions, for defense. Consequently, my investigation began with the distinct nature of human consciousness or cognition and has expanded to consider other adaptations that facilitate the behaviors that separate humans from their genetic relatives and ancestors. From consciousness or cognition and tool-making, our brains have evolved survival techniques, such as pattern-creation and -detection, to deduce cause and effect. The understanding of this relationship facilitated an enhanced curiosity and need to know the cause of virtually all events. This curiosity led, in turn, to an understanding of "agenticity," the belief that moving objects may have a living force with meaning, intention, and agency. The combination of pattern-detection and recognition of the cause-and-effect relationship fostered the "superstitious" human brain and, ultimately, religious beliefs. Thus, it was the development of enhanced consciousness and the corresponding act of tool-making that, ultimately, established the framework for higher-level concepts and belief systems. Finally, as indicated in the archaeological record, our species developed an enhanced sense of self-awareness, empathy, and a theory of mind (ToM) along with beliefs in an afterlife. ToM refers to the realization that others have thoughts, agendas, beliefs, desires, intentions, and perspectives that may or may not resemble one's own. Likewise, in humans' close genetic relatives and contemporary hunter-gatherers alike, the advanced need and capacity for "belongingness" provides the social foundation for accepting and even searching for something larger than oneself.

Great apes use stone tools, such as hammer-stones, to crack open nuts, and the hominin species that developed tool-making capabilities superior to those of the great apes must have had significantly different brains from those of other hominids. The archaeological record indicates that hand axes were widely used from 1.7 million to 100,000 ya. This is a very long time with very little apparent technological innovation. The hominins that lived during this time included *Homo habilis, Homo erectus, Neanderthals,* and, finally, *Homo sapiens.* The brains of these hominins increased in terms not only of size but of connectivity within the brain, areas of which were transformed, in some cases even at the cellular level. These changes must be taken into account when evaluating the evolution of functional capabilities because they allowed for better tools as well as improved language skills and socialization, and, eventually, the domestication of animals and grains and worship of gods. This book traces these developments over the hominin journey from makers of rudimentary stone tools to god-worshipping hominins capable of traveling to the moon.

Contents

1. The Brain and Tool-making

Only members of our genus, the *Homo family*, have the capacity to make a stone tool as sophisticated as the one that I found in Israel. Our genetic relatives, chimpanzees, use stones to break open nuts, and there are reports of chimps saving or hiding favorite stones. However, they don't seem to fashion or try to change the shape of a stone to increase its effectiveness as a tool. They may also make tools for retrieving termites from logs, shaving or breaking the unwanted branches from a long stick to facilitate its penetration of an opening in a nest. Similarly, gorillas may bend or break branches to prepare a spot for sleeping, but that is about as far as their tool-making goes. These species evidently have not been subject to genetic selection for the capacity to make more complex tools, though it is possible that they have the capability to learn the necessary skills.

Instructive in this regard is an experiment by researchers at Indiana University and Georgia State University that involved a researcher sitting with a bonobo and showing it how to chip at a stone.[8] The ape was fairly successful in imitating the human's actions but, left on its own, switched modes of production, throwing the stone against the floor or another stone to fracture it rather than employing the knapping techniques that the researcher demonstrated. The animal's failure to adopt the more efficient approach to tool-making may have been due to a physical or motor control limitation of bonobos attributable to the absence of a need to expend the effort to make this type of tool owing to the ready availability of food.

Archaeologists and paleontologists have discussed the development of stone tool-making at length since first associating stone tools with early hominins and the contextual dating of the tools. The earliest stone tools date to roughly 3.3 million years ago (mya) in the form of hammer-stones for breaking open nuts, such as chimpanzees use, as well as smashing bones for the marrow, butchering small prey, and scavenging larger game, activities for which chimpanzees do not use tools.[9] By approximately 2.5 mya, hominins were fashioning sharp cutting edges for the processing of dead animals by chipping flakes from larger stones.[10] These tools are commonly referred to as "Oldowan" after the region in eastern Africa where they were first identified. The chipping process left a distinct remnant called a "core" (Figure 2). However, there is an interesting implication in the notion that scavenging and butchering were the impetus for the initial development of tools,[11] specifically, that the fashioning of tools for the purpose with percussion methods dates as far back as

Australopithecus. To butcher a carcass with a stone flake with precision, a hominid must hold and manipulate it with an opposable thumb. This capability is not seen in other animals and may have contributed to transforming or enhancing consciousness or self-awareness. As discussed presently, the areas of the brain that facilitate precise body movements include the inferior parietal lobe, which complements the anterior cingulate and insula—structures that are related to self-awareness.[12]

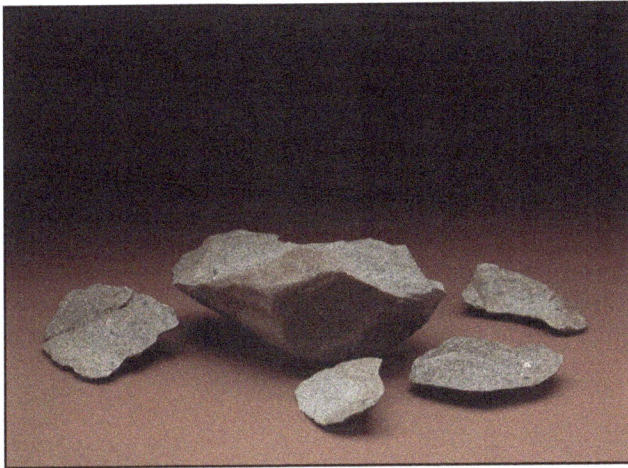

Figure 2. Stone core and flake from Lokalalei, Kenya around 2.3 mya.
Image credit: Human Origins Program, NMNH, Smithsonian Institution

As noted, the type of hand axe that I found is of the so-called Acheulean style, which came into use about 1.7 mya. *Homo habilis* roamed the savannahs of Africa from about 2.3 to 1.4 mya. Since *Homo erectus* roamed these savannahs from about 1.8 mya to 300,000 ya, the two species coexisted for some

400,000 years.[13] Thus, the Acheulean-style hand axe appeared near the end of the *Homo habilis* timeframe and remained in use for the entirety of the *Homo erectus* timeframe. It seems that *Homo erectus* and *Homo sapiens* embraced and exploited this technology because of these species' superior mental capacity. The logical assumption is that these hominins learned that the core remnant (from chipping flakes for cutting) serves as a useful hammer, club, or stabbing or throwing weapon. However, the complexity of the hand axe is far greater than that of the simple knapped flakes and hammer-stones. The primary complexity lies in the fact that the axe is bifacial, that is, worked on two sides, indicating that the maker had to think several steps in advance while understanding the impact of each strike. This kind of forethought and analysis represented a giant step for hominins.[14] The precise forces behind or causes of this transition remain unclear, but it does seem possible to chart the changes in the hominin brain that enabled this advance in tool-making.

1.1 Brain Size and Architecture

Much of the discussion about advances in the crafting of stone tools, then, concerns the evolution and development of the brain. Comparisons of *Australopithecus, Homo habilis, Homo erectus, Neanderthal,* and *Homo sapiens* skulls reveal a significant increase in brain size over time (Figure 3). Size alone, though, does not necessarily mean greater intelligence.

The brain of a gorilla, for example, is significantly larger than that of a chimpanzee, but the latter demonstrates more mental capabilities. Likewise, *Neanderthals* had larger brains than *Homo sapiens* but were unable to compete with them. Therefore, the configuration of the brain and/or its cellular structure may correspond to the significant behavioral differences among these hominins.

Species	*Australopithecus*	*H. habilis*	*H. erectus*	Neanderthal	archaic *H. sapiens*	modern *H. sapiens*
Time	4.2-1.9 mya	2.3-1.4 mya	1.8 mya-300,000 ya	230,000-40,000 ya	600,000-100,000 ya	100,000 ya-present
Brain Size	400-475 cc	630 cc	750-1,250 cc	1,450 cc	1,350 cc	1,350 cc

Figure 3. Brain size of select hominin species.

Paleo-neurologists such as E. F. Torrey study five major research categories associated with hominin brain evolution, namely, "studies of human skulls; studies of ancient artifacts; studies of postmortem brains of humans and primates; studies of brain imaging of living humans and primates; and studies of child development."[15] Neurologists have distinguished four major parts in the human brain cortex, the frontal, temporal, parietal, and occipital lobes (Figure 4).

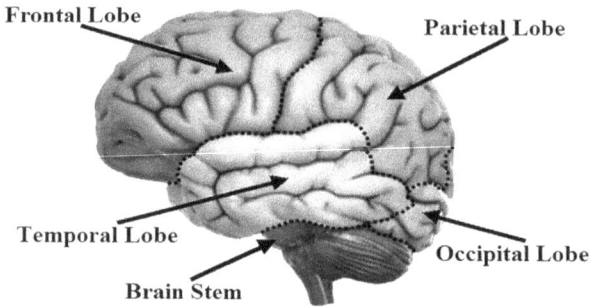

Figure 4. Lobes of the modern human brain.
Image credit: drawing by Randy Mohr.

The brain evolved from the brain stem up. That is, the oldest (paleo) part of the brain is the brain stem (also shown in Figure 4), which is present in virtually every species of mammal. Over the millennia, the various lobes developed to accomplish specific capabilities. In the hominid brain, the last development was the prefrontal cortex of the frontal lobe.[16] However, though perhaps coincidently, the brain of a child develops in the same sequence as the *Homo sapiens* brain evolved in accordance with the dictum that "ontogeny recapitulates phylogeny." Recent studies indicate that the prefrontal cortex may not fully mature until a person's mid-twenties.[17]

Modern humans' nearest living cousins as revealed by DNA analysis (i.e., the great apes) have the same major parts of the brain, as was probably the case for the ancestors of modern humans beginning with *Homo habilis.* However, the structure and relative size of the lobes differ greatly across these species. The comparison of the shapes of the skulls shown in Figure 5 makes clear that the modern human skull is not only larger but also has a larger forehead and smaller lower back region than human ancestors. The frontal lobe is located directly behind the forehead and the occipital lobe at the back of the skull. The behavior of patients with brain injuries indicates that the frontal lobe is involved in advanced reasoning and social interactions.[18]

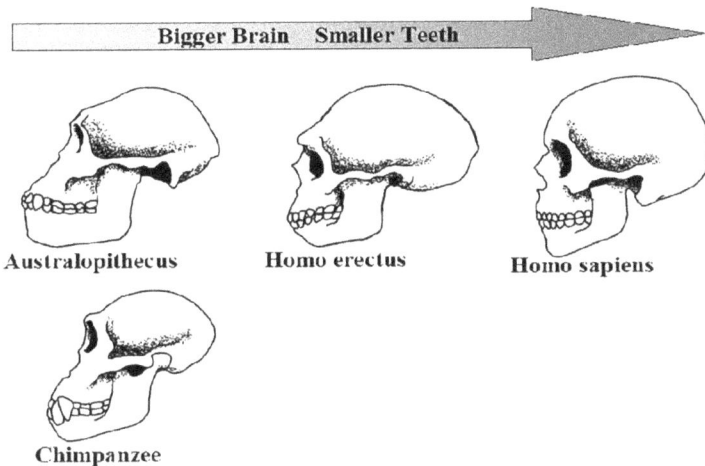

Figure 5. Evolution of skull shape.
Image credit: drawing by Randy Mohr.

The higher forehead of the *Homo erectus* skull compared with those of chimpanzees and *Australopithecus* provides for an increase in the size of the frontal and parietal lobes. The use of tools by *Homo erectus* also shows an increase in complexity or sophistication over earlier human ancestors further indicating a "smarter" and more capable species.

Further, casts of the interiors of skulls, known as endocasts, may indicate a decrease in the relative size of the occipital lobe (visual processing center) at the rear of the brain and a significant relative increase in the size of parietal, temporal, and frontal lobes, though paleo-anthropologists continue to debate this issue. The evidence for this shift is the position of the lunate sulcus, shown in Figure 6 as a crescent-shaped line of depression touching on the occipital, parietal, and temporal lobes. There is a great deal of connectivity among the three lobes, but this line seems to represent a distinct separation of functionality.[19] Chimpanzees have a lunate sulcus fairly high up in the brain, whereas that of modern humans is further back and lower down. This difference may indicate that, over the course of evolution and the development of capabilities including tool-making, cognition, and language, the parietal, temporal, and frontal lobes increased in size relative to the occipital lobe.

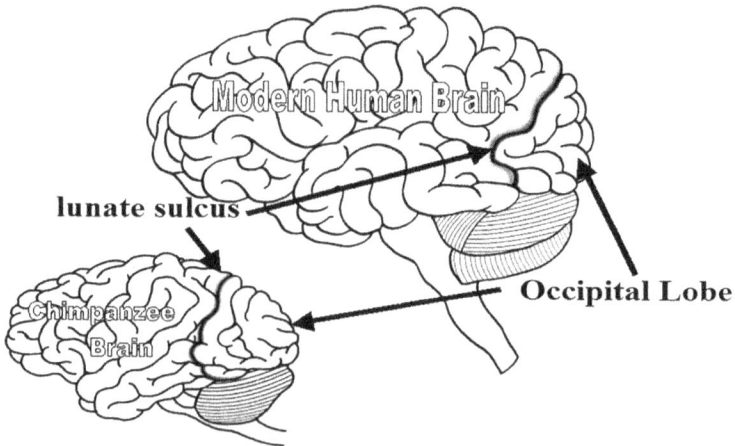

Figure 6. The lunate sulcus.
Image credit: drawing by Randy Mohr.

In his book *The Deep Understanding of Ourselves* (2019), Joseph LeDoux delineated five other distinctions between the human brain and the brains of other mammals:[20]

1) The human prefrontal cortex has distinct patterns of connectivity within and between the cell layers.

2) The interconnectivity of the neurons in the human prefrontal cortex to the neurons in the parietal and temporal cortical areas is "stronger" than that in apes.

3) The expression of genes in the human prefrontal cortex differs from that in apes "in relation to 'energy metabolism and synapse formation."

4) Mammalian cortices have five or six layers of cells. The cortices of humans and apes have six layers, one of which (layer four) contains granular cells. This thin (4-6 mm) layer covers a substrate of axons that connect the cortical cells in the various lobes of the brain.

5) The human prefrontal cortex has a frontal pole, a structure not found in the brains of apes.[21]

E. Fuller Torrey noted two additional distinctions between the human brain and those of other animals:

6) Human brains feature von Economo neurons (VENs) related to self-awareness than any other species.

7) Human brains feature a greater concentration of mirror neurons that are related to empathy and the ToM.[22]

These seven significant distinctions seem central to how and why *Homo sapiens* have progressed further than any other hominid.

Psychologists, philosophers, paleo-anthropologists, and neurologists have long tried to link tool-making with consciousness and the evolution of cognition.[23] These efforts require definitions of "consciousness" and "cognition" and raise the issue of where these faculties reside in the brain. LeDoux defined cognition as "the processes that underlie the acquisition of knowledge by creating internal representations of external events and storing them as memories that can later be used in thinking, reminiscing, musing, and when behaving."[24] The maker of a hand axe, then, would have stored the image of the tool and the actions of chipping away at a stone and recalled them to create a new axe. As noted, *Homo erectus* pioneered the Acheulean hand axe, so the brain of this species must have been the first with sufficient cognition and working memory or "mental workspace" to produce such an axe.

Also as noted, the technology of the Acheulean hand axe appears to have stagnated for about 1.35 million years except in some areas in eastern Africa where the environment fluctuated.[25] Rick Potts, a paleontologist with the Smithsonian Institution, reported in 2020 the discovery of alternating periods of resource scarcity from 500,000 to 320,000 ya in one part of the Rift Valley in southern Kenya.[26] This change from a relatively stable environment was accompanied by a change in the flora from short to long grasses and in the fauna from large-bodied herbivores to smaller-bodied mammals with diverse diets. Most important for the present discussion, the diversity of stone tools, including spear points, increased. Potts and his team proposed that hominids in this area evolved to handle frequent climate and environmental shifts, which evolution could have resulted in increased brain plasticity.

Plasticity refers to the ability of the brain to change, modify, and, essentially, rewire itself to accommodate both internal changes such as those resulting from injury and external changes in the environment. However, as noted, this diversity of tools has been documented only in this limited area while hominins in other parts of Africa continued to use the same hand-axe technology with no significant improvements or even variations. *Homo erectus* apparently invented these hand axes, and they remained the primary technology of archaic *Homo sapiens* for thousands of years. Something began to change approximately 100,000 ya, however, across a broad geographic area, with the advent of "early *Homo sapiens*," for a variety of tools began to appear in the archaeological record, including sophisticated forms made of bone.[27]

1.2 Working Memory, Mental Workspace, Cognition, Consciousness, Imagination, and Self-awareness

Research on the development of the brain can be confusing with regard to the semantics and concepts involved. Experts in several disciplines have argued that consciousness is a prerequisite for tool-making and is primarily the domain of *Homo sapiens* or modern humans. Further, the terms "working memory" and "mental workspace" reflect attempts to identify the internal neural systems or areas of the brain responsible for the imagination, dreams, and consciousness or cognition.

These concepts are as elusive, however, as is their location in the brain. In part to explore such issues, the sub-discipline of neurology known as 'cognitive science' developed. Cognitive scientists have shown that consciousness exists in the animal kingdom, including humans, as a matter of degree or magnitude. That is, other species experience various forms of consciousness, imagination, and even dreams. Thus Darwin stated in *The Descent of Man* (1871) that "The difference in mind between man and the higher animals, great as it is, certainly is one of degree and not of kind."[28]

Consciousness requires at least a rudimentary cerebral cortex.[29] In fact, extensive regions of the cortex may be removed or injured without rendering an individual unconscious. It follows that any animal with a cortex experiences some degree of consciousness. In hominids, individuals do not become unconscious except in the case of a disabling event to the area of the brain called the ascending reticular activating system (ARAS). Damage to other areas of the cortex, by contrast, may render an individual cognitively impaired but not unconscious.[30] However, the ARAS alone is not responsible for consciousness. Rather, this system receives input from many areas of the brain to provide or create the subjective experience of consciousness.[31] Consciousness, then, has been well described as "a bubbling cauldron of subjective sensations, perceptions, memories, and feelings, all competing for our attention."[32]

For the purposes of this book, I distinguish the concepts of consciousness, working memory, mental workspace, imagination, and self-awareness. Consciousness refers to the overall executive function that facilitates the performance of multiple functions concurrently or in sequence. Working memory, mental workspace, and imagination utilize specific areas of the brain that serve temporarily to perform specific tasks. These terms are often interchangeable in terms of describing the faculties necessary to make stone tools. Making stone tools also requires, however, a point of reference, that is, the self.

1.2.1 The Self

The development of advanced tool-making may not have depended as much on consciousness as on an enhanced degree of "self-awareness." One of the major differences between the brains of humans and those of great ape brains just mentioned is the increased presence of special neurons (VENs) that appear to be critical for self-awareness[33] and reside in the anterior cingulate, anterior insula, and to a lesser degree, lateral prefrontal cortex (Figure 7). These VENs, also known as "spindle neurons," constitute only 1-2% of the total neurons in the cortex but are about four times the size of other neurons. The brains of some other species have VENs but in lower proportions than they are found in human cortex.[34] These observations suggest that consciousness and self-awareness are present in many species, again, as a matter of degree or on a continuum.

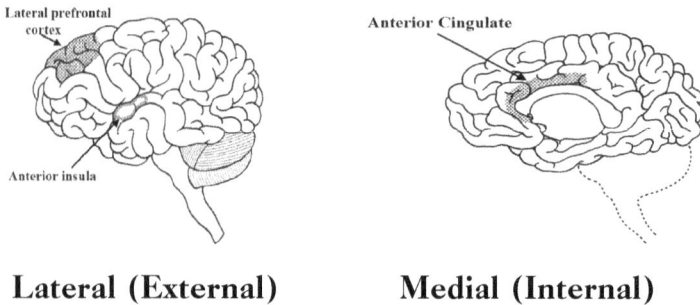

Lateral (External)   Medial (Internal)

Figure 7. Locations of von Economo neurons (VENs) in the human brain. *Image credit: drawing by Randy Mohr.*

Torrey states further that this assemblage of a large brain, neural network, enhanced parietal lobe, and "working space or memory" provides for the cognitive ability to hold and manipulate objects, including the two stones used to produce a hand axe, with great precision, making it an indicator of "self-awareness."[35]   In this case, self-awareness involves knowing that the "I" that is making the axe is separate from the tools used to produce the axe; it is a realization that "I" know what "I" am thinking about and can, at will, maintain or shift "my" attention. Thus, Torrey called *Homo erectus* the "aware self." Making a hand axe, then, represents a huge step along the continuum of self-awareness, as discussed further in the section on working memory (2.1.3).

As thus described, the "self" must willingly maintain its attention. This statement, though, begs the question of what "attention" is and where it came from. Michael Gazzaniga, a pioneer in neuroscience research, has defined attention as "selective signal enhancement," a process for managing the enormous amount of sensory input that the body sends to the brain.[36] He distinguished top-down attention and bottom-up attention. Top-down attention occurs when an individual intentionally selects a target for conscious consideration, such as when searching for something or making a stone tool. Bottom-up attention occurs in reaction to external stimulation.[37] Stephen Hinshaw, on the other hand, has distinguished five types of attention: automatic, in response to a new or unexpected stimulus; selective, involving a choice to focus on an object or event; sustained, involving focus on the completion of a task (such as a lecture or making a stone tool); executive, involving focus of the executive function on planning, over-riding distracting stimuli, and working through errors; and working, involving the simultaneous deployment of multiple skills (such as mathematics and physical dexterity) that require significant mental capacity.[38]

Hominins that may be considered self-aware and possessing enhanced attention include *Homo erectus,* archaic *Homo sapiens, Neanderthals*, and modern *Homo sapiens.* Naturally, the degree of self-awareness varied, increasing in intensity or capability as the brain evolved.[39] According to Torrey, *Homo erectus* "would have been able to admire its own reflection in a still pool," but *Homo sapiens* is the first species of hominins able to understand the thinking of others of its kind.

2. Memory-making, Schemas, and Patternicity

The Making of Memories

When the body's receptors (eyes, ears, skin, etc.) sense something, the temporal and parietal cortical areas of the brain are the first to <u>perceive</u> it.[40] As neurologists and psychologists like to say, we see with our brains. So, while the cortex provides subjective perceptions of things, the sense organs send signals to the brain that, ultimately, the cortex processes as an image, sound, or feeling of texture. Each of the senses provides only a piece of the actual external object or stimulus; the brain, in particular, the cortex, fills in the missing pieces based on inference.[41] Thus neurologists also like to speak of the "inferential brain." The most common, and best, example of the inferential brain at work is in the processing of visual information (Figure 8).

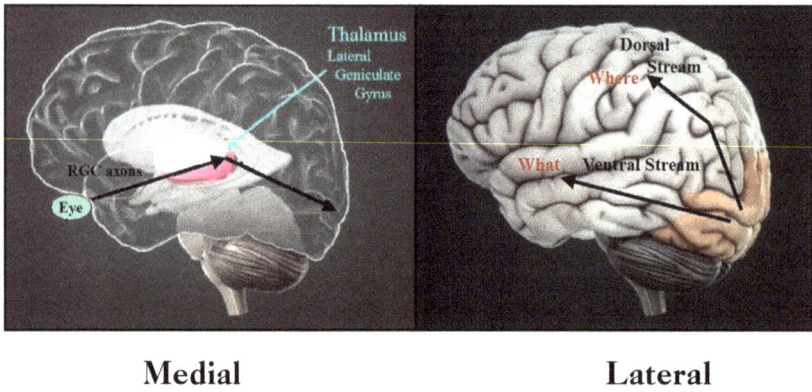

Medial   **Lateral**

Figure 8. The processing of visual information in the brain.
Image credit: Cold Spring Harbor Laboratory DNA Learning Center 3D Brain app
(Arrows and red and black labels added by the author).

Processing is necessary because it is impossible for the brain to receive and process all of the images or data that the eyes receive. Billions of photons strike the retinas, where they are captured by two types of photo receptors: rods, numbering around 120 million, which are the main receptors for low light conditions, and cones, numbering around 6 million, which are the main receptors for high light conditions and color. However, only 1 million retinal ganglion cells (RGCs) travel from the eye to the lateral geniculate nucleus (LGN) in the thalamus for further processing.[42] Consequently, there are not enough neurons or pathways to process and/or store all of the photon stimuli that the eyes receive. The RGCs attempt to compensate by rapid repetitive firing but still cannot detect and transmit the signals from all of the photons entering the eye, so some of the data are lost. I emphasize this point because it is central to the concept of pattern completion discussed later in this chapter.[43]   From the LGN, neurons carry the information to the occipital lobe, shown in orange in Figure 8, where it is projected onto a retinotopic map. The subjective experience or perception of these data requires their transfer to the cortex. Again, as shown in Figure 8, within the occipital lobe, various types of information are sent to distinct destinations. Information relating to the position and velocity of objects travels to the visual cortical areas in the parietal lobe in what is called the dorsal stream. Information about what something is travels to the temporal visual cortical areas in the ventral stream.[44] In the process of transit to the various areas of the cortex, some data may be lost or misdirected.

Within the cortices, pieces of related information about a subject or event are bundled into what neurologists and psychologists call "schemas," which are mental models or patterns.[45] These schemas then serve to generate expectations about objects, events, and situations. For example, the temporal visual cortex contains neurons that only fire when a face is perceived—in fact, some of these neurons only fire when a specific face, such as a person's mother, is perceived. Schemas are the foundation or building blocks of cognition,[46] for the bundling of pieces of information allows the executive control function to complete patterns involving other pieces of information so as to, for example, put together the fragments of an image or recall a song based on a few notes. This processing enhances the ability to connect events and view a life as a unified story rather than a string of isolated experiences. Cognition has long been considered a prerequisite for sophisticated tool-making. Since the schemas just described provide the building blocks of cognition, they must facilitate sophisticated tool-making among other complex tasks.

2.1 Types of Memories

Having considered the transformation of sensory data into subjective conscious experiences in the cortex, the next issue regarding the brain and tool-making is how and where these experiences are stored and retrieved for future reference. Neurologists and psychologists have identified many types of memory, of which, for the purposes of this discussion, the focus is on four: episodic, semantic, working, and autobiographical. Episodic and semantic memories were probably the first to develop since most higher animal species possess these capabilities to some extent. However, an enhanced working memory, which is either essential for consciousness or perhaps equivalent to consciousness itself, seems to be the domain of hominins alone, having first appeared in *Homo erectus*. Autobiographical memory appears to be a later development that is now found only in *Homo sapiens.*[47]

2.1.1 Episodic Memory-making

An episodic memory is simply an episode or event that

an individual experiences. It is an event or scene that contains visual objects, perhaps including other individuals, other sensory impressions (such as smells and sounds), and the self in the context of other things, for example, spatial data. This type of memory is associated with the dorsal stream, which is shown on the top right in Figure 8. Significant attributes of an episodic memory are the inclusion of the self and emotional feelings of ownership that are time stamped and contribute also to autobiographical memories.[48]

Data from an episode pass from the eyes to the LGN in the temporal cortex and then to the visual cortex in the occipital lobe, the parietal and temporal visual cortex, the parahippocampal cortex, and, finally, to the hippocampus and amygdala for storage (Figure 9).[49] The hippocampus is part of the brain's limbic system, which is also the center of emotions. The brain is continually monitoring an individual's physical and emotional states, and these states form part of the episodic memory. Consequently, episodic memories are often associated with emotions. This is a critical point because, as will be seen, the associated emotions are required for subsequent judgments and decision-making. An episode persists for a while but may eventually dissipate into other parts of the cortical tissue. A perceived object becomes transformed into a schema or semantic memory, and it seems that the context of an episode can likewise dissipate and be lost. However, episodes that involve other people, significant objects, intense emotions, or particularly abnormal or stressful contexts may also be retained in the hippocampus and amygdala along with their emotional schema.[50]

2.1.2 Semantic Memory-making

Semantic memories concern facts about the world

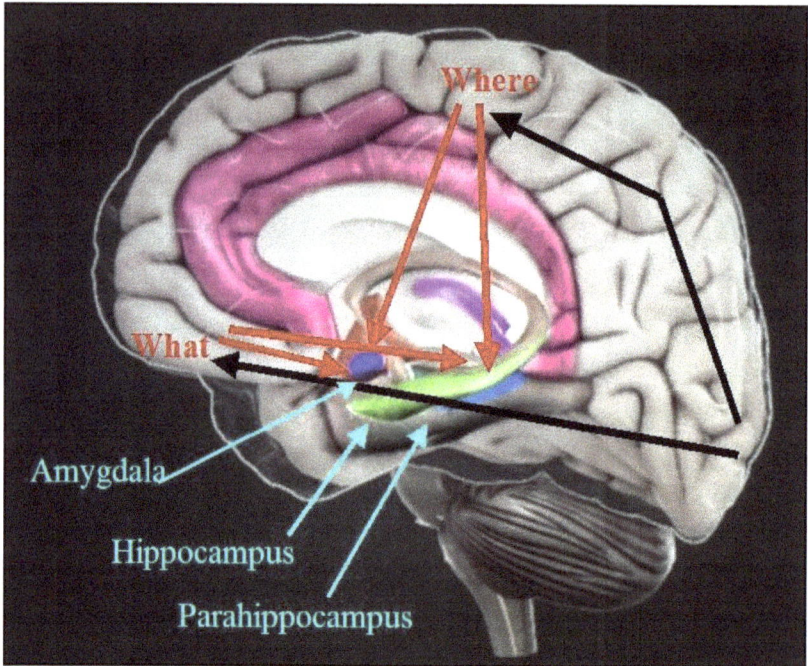

Figure 9. Processing of an episode in the brain.
*Image credit: Cold Spring Harbor Laboratory DNA Learning Center 3D Brain app
(Red and black arrows and red labels added by the author).*

and objects other than the self that become or are schemas, models, or patterns.[51] Such memories may include the self when first encountered as episodic memories, but the self and the context of the schema are usually lost over time. Thus, as shown in Figure 10, the context, or "where," dissipates over time, as indicated by the dotted lines. For example, an individual may retain a schema of a pencil but not necessarily his or her first encounter with a pencil (unless it was traumatic or otherwise significant). Though the original memory may have been episodic and resided in the hippocampus, it remains primarily in the lateral cortex or migrates there. A semantic memory, therefore, is not necessarily attached to an emotional schema and often only resides in the cortex.[52]

The schema or pattern for an item or event, or "what," is stored in the lateral cortical area of memory and retrieved whenever an individual encounters the same object or a

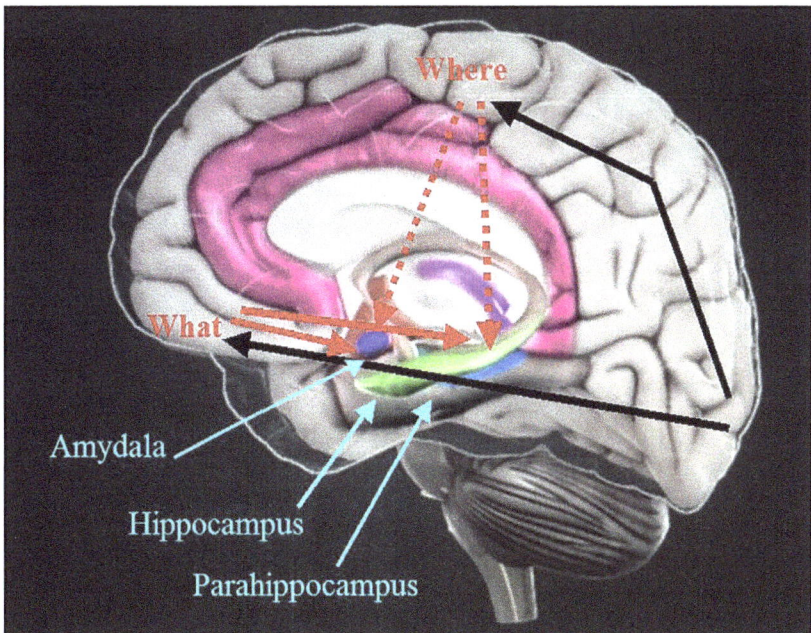

Figure 10. Dissipation of context over time in a semantic memory. *Image credit: Cold Spring Harbor Laboratory DNA Learning Center 3D Brain app (Red and black arrows and red labels added by the author).*

similar one.[53]   Consequently, individuals' preconceived images of objects from memory encounter newly perceived images in real time. What is seen or experienced in any situation depends on, and is influenced by, what has been seen or experienced in the past.[54] Some memory neurons in the cortex are very specific, such as those mentioned earlier that fire only in response to the pattern of a face. These facial recognition neurons are precise and selective in "cognizing" a new face or object into patterns. The sight of a face or object that provides a pattern already in the memory results in "re-cognition" of the pattern. When a new experience does not match the expectations in the memory, either the previously stored schema of the event or object assimilate and update the new information or the new experience creates an unsettling sensation of surprise or dissonance. In the latter case, the new information then produces an episodic memory because it is linked to a single novel episode. The episodic memory, then, both assimilates similar information encountered at a later time and creates new schemas, or patterns, through generalization.

2.1.3 Working Memory

An enhanced or expanded working memory and with it "self-awareness" appear to have became evident in the archaeological record with Homo erectus about 1.8 mya.[55] Perhaps the most significant indication of this transformation was the development of Acheulean hand axes such as the one that I found. As noted, this is a bifacial tool, the crafting of which requires the maker to hold an image (of the tool) in the working memory while holding a stone in each hand and calculating precise, forceful strikes at specific angles. In other words, this task involves the temporary storage of information (i.e., images) related to a specific task in the mental workspace or working memory.

Until very recently, little empirical data was available to support discussions of working memory, and researchers had to depend on psychological and philosophical theorizing and deductive reasoning. Technological advancements with applications in psychology and neuroscience have now made it possible to detect the areas of the brain that are active during specific tasks, in particular, functional magnetic resonance imaging (fMRI), the technique introduced earlier that involves the use of strong magnetic fields to detect energy or oxygen in active areas of the brain. Equipped with these tools, researchers have focused on mental workspace and working memory in relation to mental activities and the manipulation of objects. Their findings suggest that the mental workspace facilitates imagination. In an effort to locate it, several recent neuroscience studies have looked at mental representations (i.e., visualization) rather than operations and found that visual perception,[56] visual imagery,[57] and dreaming[58] take place in the visual cortex or occipital lobe in coordination with other areas in the brain that, together, can be described as "imagination."

Especially noteworthy in this context is a 2013 study by scientists from Dartmouth College who used fMRI to detect changes in blood oxygenation level-dependent (BOLD) data across the active areas of the brain while test subjects maintained and manipulated visual imagery.⁵⁹ The resulting fMRI of a brain while manipulating visual imagery is shown in Figure 11.

The researchers described their data as "revealing a widespread cortical and subcortical network that operates on visual representations in the mental workspace. This network

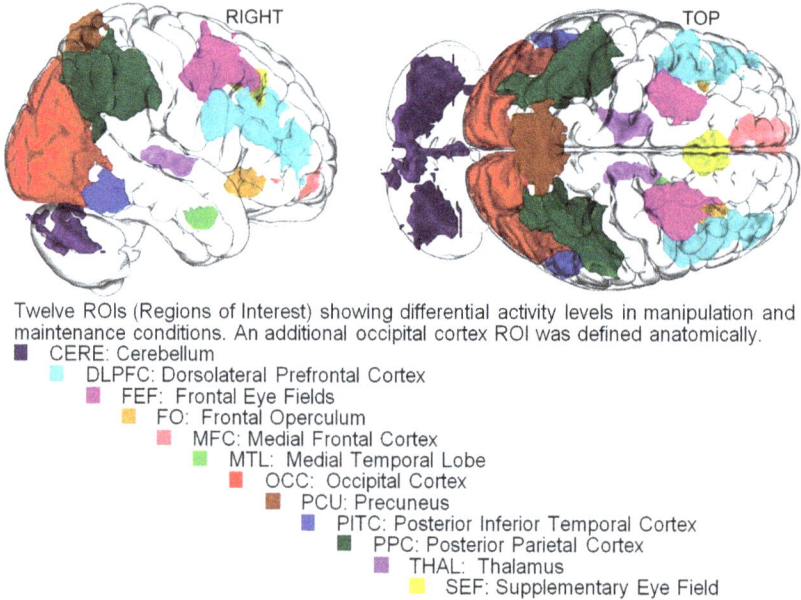

Twelve ROIs (Regions of Interest) showing differential activity levels in manipulation and maintenance conditions. An additional occipital cortex ROI was defined anatomically.

■ CERE: Cerebellum
■ DLPFC: Dorsolateral Prefrontal Cortex
■ FEF: Frontal Eye Fields
■ FO: Frontal Operculum
■ MFC: Medial Frontal Cortex
■ MTL: Medial Temporal Lobe
■ OCC: Occipital Cortex
■ PCU: Precuneus
■ PITC: Posterior Inferior Temporal Cortex
■ PPC: Posterior Parietal Cortex
■ THAL: Thalamus
■ SEF: Supplementary Eye Field

Figure 11. Active regions of the brain.
Image credit: reproduced with permission of Alexander Schlegel.

includes four core regions spanning the DLPFC (dorsolateral prefrontal cortex), PPC (posterior parietal cortex), posterior PCU (precuneus), and OOC (occipital cortex) that manipulate the contents of visual imagery. The DLPFC appears to be part of a network that maintains representations in working memory via attention. Our data provide empirical support for this model by showing that the DLPFC and PPC mediate not only the maintenance of representations in working memory but also the manipulation of those representations. Thus, these areas may form the core of a system that mediates conscious operations on mental representations, in this case the contents of visual imagery represented at least partially in the occipital cortex."[60]

Imagining a stone tool and employing that imagination, or working memory, to create the stone tool, however, is even more complex than these results suggest, and research into tool-making using fMRI also has shed light on what goes on in the brain during the actual crafting of, specifically, an Acheulean hand axe. These fMRIs show a great deal of activity in the four core regions of working memory described above as well as the left superior frontal gyrus (LSFG), which is important in terms of "contributing to information monitoring and manipulation . . . between default and control networks during internally focused, goal directed cognition, and particularly the planning/simulation of future actions."[61] This "mental time travel," limited as it may be in tool-making, is critical for autobiographical memory.

It seems that making something like a hand axe involves, not copying a specific axe, but rather retrieving a generic image of an axe from memory that resides in the hippocampus, amygdala, and/or cortical areas, including the "what" lateral cortical area, and moving the image into the working memory.

This working memory includes the dorsolateral prefrontal cortex (DLPFC), posterior parietal cortex (PPC), and the occipital cortex (OCC).[62] Again, I consider this visualization to be equivalent to "the imagination." *Homo sapiens* are the only species that certainly possesses this enhanced working memory or imagination, which facilitates not only tool-making but also dreaming, the recollection of images from the past, and forming a vivid picture of the future. It seems that most animals also dream, as many people have observed in dogs. In fact, as mentioned, all animals with a cortex seem to dream.[63] Lower animals may have limited working memory or imagination, but they do not, of course, use it for the advanced tool-making in which humans engage.

Researchers have attributed to *Homo habilis* (literally "handy person") the extensive use of the Oldowan tool assemblage, in particular, chips and flakes for scavenging and butchering animals. To *Homo erectus* they have attributed the development of the Acheulean-style hand axe of the type that I found. This change in tools reflected a change in food-acquisition strategies, mainly from scavenging to hunting, presumably reflecting, in turn, differences between the brains of *Homo erectus* and *Homo habilis*. A 2015 fMRI study of test subjects who had been trained to make both Oldowan and Acheulean tools to assess differences in task difficulty, learning curves, and brain use and development found that the Oldowan tools did not require the precise prediction and evaluation that flake-production required.[64] Avoiding mistakes was not crucial for the production of flakes to be used for butchering, but, according to the authors of the study, "explicit prediction and evaluation of tool-making action outcomes . . . is a normal part of Acheulean hand axe-making skill." Further, the cognitive control demands of Acheulean "tool-making are modulated by a combination of tasks, training, and technology." The executive control function of the working memory is associated mainly with the mid-DLPFC and also with a broad network of relevant regions. The LSFG contributes to the strategic evaluation of actions and serves as the region's primary node for interactions between default and control networks during the focused planning of future actions, such as the axe-maker's next strike. This, as has been seen, implies at least rudimentary "mental time-travel" capability.

In the case of the hand axe, the enhanced neural network facilitating working space, or the enhanced imagination, allowed our human ancestors to imagine possible courses of action (e.g., kill a deer), recall the best tool for the job (a hand axe), and upload the image of that tool from the DLPFC and posterior precuneus into the occipital lobe. Then, holding that image in the working memory or imagination while coordinating messages from the cortex to the motor areas (premotor cortex, supplementary motor area, primary motor cortex, and cerebellum), the maker of the axe performs the appropriate hand motions to chip away at a rock to form a tool resembling the image held in the imagination. Modern humans' genetic cousins, the great apes, cannot create tools of this complexity. The hominins in which this enhanced capability evolved experienced and survived selective environmental pressures that other hominins, and the great apes, did not experience presumably because their habitats were more stable.

These considerations raise the question of how the various areas of the brain work together in the manner just described, in particular, which part "runs the show" and which parts provide the data necessary for it to do so. The answer is not entirely clear, but researchers have a decent grasp of the hierarchy of events. In the performance of a given task such as making a stone tool, the relevant data, whether acquired in real time or retrieved from memory, are held in the mental workspace by an "executive control function." This function may use the data for reference and manipulate them during the performance of the task, which is to say, pay attention to them.[65] The fMRI results strongly suggest that the DLPFC is the part of the network that maintains representations in the working memory and, in effect, maintains attention. Thus, the DLPFC, as the executive control function, summons data from memory that resides in such places as the lateral or dorsal cortex, hippocampus, and amygdala to temporary storage areas. However, the working memory seems insufficient to store an entire set of images or events related to the performance of most tasks.[66] Consequently, the brain—and, probably, the DLPFC specifically—accepts or selects only pieces of the images or events that have been retrieved. This selective acceptance is a critical point in my larger argument.

Within this system, then, there are at least five places where data may be lost or conflated:

1) in the initial processing of the incoming data (In this case, there are not enough RGC axon neurons to capture all of

the photons that strike the retinas and map the signals in the LGN.),

2) in transit from the LGN to the occipital lobe,

3) in transit from the occipital lobe to the lateral and parietal visual cortices,

4) in transit from the visual and parietal visual cortices to the parahippocampus, hippocampus, and amygdala, and

5) during the retrieval of a memory or schema and its transit to the working memory.

As a result, only fragments of the original image or experience remain in an individual's consciousness after he or she sees something, for the image or experience initiates a search for a pattern either in it or that can be used to supplement gaps in the fragments.

The human brain, then, has evolved through natural selection—or, rather, selective reproduction—to search for and complete patterns in data, that is, to "fill in the blanks" by inferring the missing pieces. As noted, neurologists speak of the "inferential brain" that rewards itself for completing a pattern with dopamine. One of the brain's primary reward systems, dopamine is associated with the nucleus accumbens (NAcc), a module or bundle of neurons in the center of the brain associated with pleasure and pain.[67] Thus, the release of dopamine in this area is also associated with the "high" provided by, for instance, cocaine as well as orgasm.[68] So also, after finishing an activity or identifying a pattern, the perception of success triggers a release of dopamine as a reward that, in turn, encourages the search for more patterns to complete.[69]

Because the executive function of working memory (again, the mental workspace or imagination) retrieves schemas from memory to process incoming visual data, the brain searches for patterns in every sense perception. We can't help it: our brains evolved to look for patterns in everything and every situation because doing so proved to be a very successful survival strategy. When it encounters new data, the executive function of working memory searches for a schema that fits the data regardless of its completeness. The working memory serves to form a pattern from the pieces of information that it receives and retrieves. The capacity to recognize faces is highly advanced because of the need to distinguish friend from foe, tribal member from non-member, and so on. Indeed, this capacity is so advanced that some individuals see the face of Jesus in a piece of toast, the Man in the Moon, or a visage in a photograph of the hills of Mars. M. Shermer termed this capability or tendency to find meaningful patterns in both meaningful and meaningless data "patternicity,"[70] and it is on display in the constant search for faces in every situation, conscious or not. Many animals can recognize faces but the early hominins who could excel at identifying the face of a foe lurking in a bush had an advantage over those who could not. A mistake in this regard that involved seeing a face in the bush when one was not actually present (i.e., a false positive) did not cost much, whereas a mistake that involved not seeing a face had the potential to remove an individual from the gene pool. As Shermer put it, "We are the descendants of those who were most successful at finding patterns."[71] Presumably, the same is true for most animals.

The compression of data during humans' neural processing is responsible for the compulsion to "fill in the blanks" to create something familiar. Figure 12 presents an example of this phenomenon in the form of patches of light and dark that do not form a complete picture. Most viewers make out a leopard lying in the grass at the center right, but, looking closer, many can also see a second animal (compare the original photo in the Appendix).

Another example is the night sky. Humans have long perceived patterns in the stars, some of which came to be recognized as constellations. More recently, some observers of

Figure 12. Indistinct image of a leopard.
Image credit: photo by the author; Photo prepared by Randy Mohr.

the sky have claimed "to see cosmologist Stephen Hawking's initials written in fluctuations in the cosmic microwave background, the oldest light in the universe."[72] Likewise, Impressionist painters such as Monet and Renoir exploited this phenomenon in their art. This drive to fill in the blanks is not confined to visual objects but is observable in music as well as series of events; as alluded to earlier, from just a few notes, it is often possible to identify a song and the artist associated with it. This aspect of the human drive for survival also creates feelings of disquiet, dissonance, or even distress when it proves impossible to complete a pattern, whether in the context of a crossword puzzle, a story, or a game. When people have a question, they want a satisfying answer and either continue searching for or create one.

This tendency, I argue, underlies the proliferation of conspiracy theories. Simply put, these theories allow people to complete patterns and answer questions. Provided with only a few data points, facts, or events, they are ready to fill in the blanks and entertain "what if" scenarios. The fewer the facts, the more extravagant the theories. Particularly interesting in this regard are the results of the experiments that Shermer performed to explore the consequences of pattern recognition, which revealed that test subjects who excelled in identifying patterns in random data also showed a propensity for belief in a religion and/or conspiracy theories.[73][74]

2.1.3.1 Action Sequencing and Ritual

Another type of patternicity was identified in early research in psychology that I call the patternicity of "action sequencing." One of the best examples of this kind of patterning is a famous experiment on pigeons that B.F. Skinner performed in the past century.[75] The pigeons learned to push a lever to receive a pellet of food, and it turned out that their behavior immediately before pushing the lever, such as turning clockwise or counterclockwise, was connected to their success in remembering how to obtain the reward. In other words, associating the events or motions leading up to success established a schema or pattern that the pigeons retained. This action sequencing is the basis of a ritual, that is, a specific action that must be performed before and/or after another action to achieve the desired outcome. When a ritual is executed in the company of others, its reinforcement is amplified as a form of "meaning-making." Many individuals employ these types of rituals, for instance "knocking on wood" or throwing salt over a shoulder after spilling some to prevent a subsequent adverse outcome. The propensity to create and execute such rituals is a cornerstone of religion, for most religions enjoin their followers to perform action sequences in a specific order at specific times.

2.1.3.2 Pattern Completion and Closure

Closure is another form of pattern completion that can

lead people astray and even into the realm of the absurd. The drive to mitigate the dissonance caused by the perception of unsolved mysteries leaves people susceptible to acceptance of and belief in notions that are counter to logic.[76] So it is that conspiracy theories have run rampant in every generation. Most are created with some kernel of fact, offering alternative, and, often, nefarious explanations for events with "what-if" scenarios to fill in missing data.

The unchecked need for pattern completion can, therefore, be detrimental when large numbers of people do so using circumstantial or false data. A widely discussed recent example of this outcome is the conspiracy movement known as QAnon referred to in the title of this book. In this case, adherents believe in the notion that a person (Q) is secretly alerting the public to the danger posed by a shadowy organization that wields power behind the scenes. I discuss this phenomenon in the context of agenticity in Section 3.1.

2.1.4 Autobiographical Memory

The fourth type of memory, autobiographical memory creates a running history of an individual's life and facilitates dreaming and imagination. In this form of memory, the processes of cognition and recognition are reversed in that data from the hippocampus, amygdala, and visual cortex are uploaded into working memory. Autobiographical memory is an extension of episodic memory in that it includes the self, location, and timestamps, that is, information with context. Thus, as shown in Figure 13, the context of an episode is loaded into and stored in the amygdala, hippocampus, and parahippocampus. Autobiographical memory extends episodic memory by stringing episodes together to create a kind of "movie" of a life, but it includes, in addition to the self and what has happened in the story so far, imagining or projection of the self into the future.[77] There are obviously no memories of the future, but a brain can, with the right wiring, string together pieces of episodic memory to create scenarios that could happen in the future. Thus, for example, by remembering the migratory patterns of prey animals, humans can envision driving them off of cliffs or into traps. This type of memory, involving the understanding or conception not only of a specific time but also of sequences of events over time, probably came to full capability relatively recently, around 40,000 ya, in *Homo sapiens.*[78]

In other words, ours may be the only species that has the ability to imagine or conceive of a long-term future, for which we have autobiographical memory to thank. Some readers may recall the scene in Lewis Carroll's *Through the Looking Glass* when the White Queen explains to Alice that "memory works both ways."

"I'm sure mine only works one way." Alice remarked. "I can't remember things before they happen."

"It's a poor sort of memory that only works backwards," the Queen remarked.

"What sort of things do you remember best?" Alice ventured to ask.

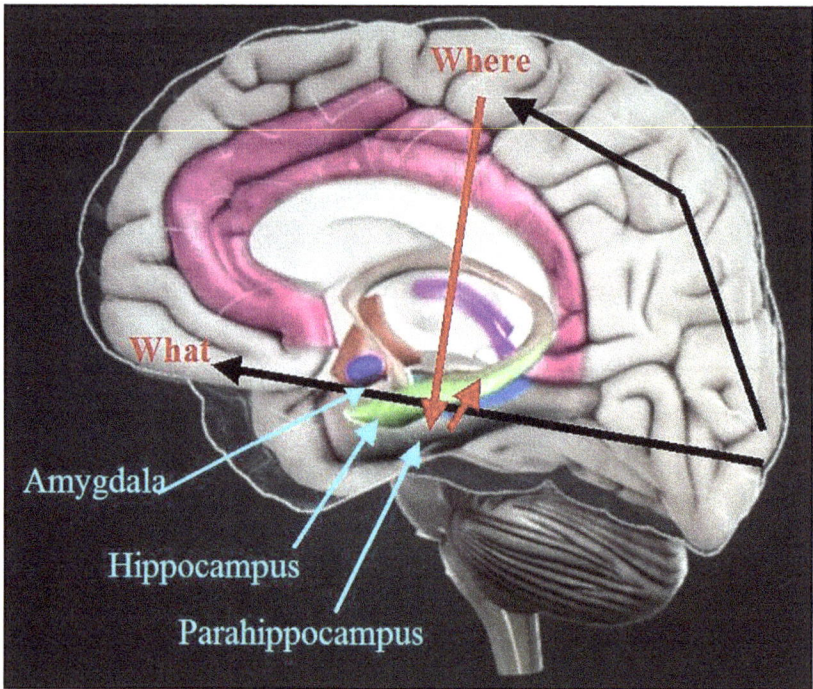

Figure 13. Parts of the brain involved in autobiographical memory.
Image credit: Cold Spring Harbor Laboratory DNA Learning Center 3D Brain app (Red and black arrows and red labels added by the author).

"Oh things that happened the week after next," the Queen replied in a careless tone.[29]

Remembering the past and imagining the future give meaning to people's lives. Complex planning for the future is the domain of *Homo sapiens.* Butterflies may migrate, and squirrels may hide nuts, but it is unclear whether such behaviors reflect conscious planning or instinct. Chimpanzees seem also to employ some planning when hunting, but it is nowhere near as complex as the planning that hominins display. Once again, this capability exists across a gradient or spectrum in various species. The working memory, autobiographical memory, or imagination allows *Homo sapiens* to craft complex hunting strategies, build complex traps with a conscious purpose, and even travel to the moon.

Notably, time may be suspended, compressed, conflated, slowed, or accelerated in the imagination. The mind may replay a car accident or other horrific event over and over again in slow motion. More positively, a new parent can also imagine her own deceased parents playing with their grandchild.

Humans' mental workspace or enhanced imaginative capabilities are absolutely amazing. Imagine an elephant dancing on the head of a pin: it happens almost instantly upon reading this sentence. To create the image, a reader uploads the images of an elephant and a pin from memory that reside in the hippocampus, cortex, and/or amygdala and then adds the motion of the elephant, even without ever having seen an elephant dance. Also, it is necessary to suspend or distort proportion given the difference in size between an elephant and a pin.

However, the imagination, in concert with the memory, can also conflate or corrupt events. For example, a colleague once insisted that I had attended a meeting at a certain place and time when I know for a fact that I was in a different country at that precise moment. She had conflated the memory of a previous meeting with a more recent one in the same setting. Sometimes, therefore, we can't, or shouldn't, believe what we remember. Likewise, the accuracy, or rather inaccuracy, of memories has been well documented in studies of eye-witness accounts of events.[80]

The ability to imagine the future also puts humans in touch with the concept of death and reveals that our own days are not infinite. That is, autobiographical memory leads to an imagined future that includes the death of the self. There are not many alternatives to this future. One is provided by the concept of mind-body dualism, with the mind being separate from the body and able to survive its death. As discussed presently, this view may have received support from the experience of dreaming and the practice of ancestor-worship. The notion of life after death brings with it the notion of immortality as well as questions about the purpose of human life generally or a human life in particular. The autobiographical memory of the incomplete series of events in a person's life—the product of patternicity, the need to complete the story—and enhanced imagination encourage thinking in terms of a beginning and end to existence. Entire religions have arisen, in part, to address these issues. Thus, E. O. Wilson hypothesized that early humans needed a story of everything important that happened to them, because the conscious mind cannot work without stories and explanations of its own meaning. The best, the *only* way our forebears could manage to explain existence itself was by developing a creation myth.[81] (emphasis in original).

2.2 Language

Another aspect of patterns is the evolution of language. Like complex tool-making, complex verbal language is a capability that is solely the domain of humans. Other species communicate through body language and a relatively limited range of sounds. Birds whistle to one another, dogs bark and whine, coyotes howl, lions roar, whales sing, and so on. However, these sounds are limited in tonal range as well as vocabulary and grammar. Some species, including gorillas, chimpanzees, and bonobos, have sets of gestures that allow for somewhat more complex communication. In fact, several scholars have proposed that gestures, as a form of sign language, were the precursors to verbal language in communicating action. Researchers at Emory University performed fMRIs on 27 great apes and showed brain asymmetry and activity in the part of the brain known as Broca's area when the animals gestured and vocalized.[82] Broca's area is associated with the generation of language in humans (Figure 14), with deaf people likewise showing activity in this part of the brain while communicating through signing.[83] Actions, expressed by verbs, are a large part of all languages and some of the first words learned by children and language students.

Language and tool-making share several key aspects. To begin with, both manual dexterity and vocal skills depend on sequence patterns in the brain, specifically in the frontal lobes.[84] These small, flexible modules can be rearranged in

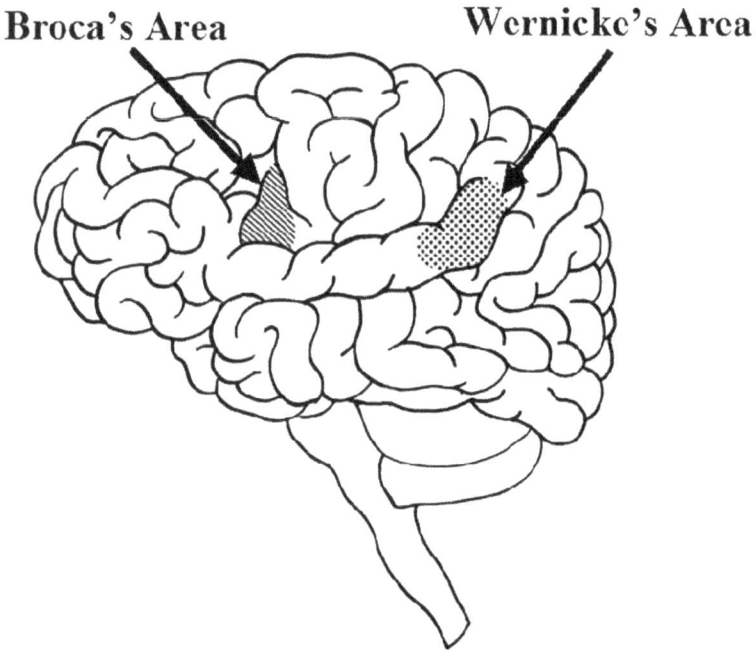

Figure 14. Parts of the brain associated with language.
Image credit: drawing by Randy Mohr.

complex ways depending on the task at hand. Similarly, crafting a hand axe or a knife requires many of the same minute motor skills, but the skills may be rearranged to obtain various outcomes. Thus, by way of further example, different sequences communicate the actions of running and walking.

The brain is asymmetrical in that there are specific areas in each half of it that have evolved to perform distinct functions. Neurologists have identified two specific areas on the left side of the brain that facilitate language processing in hominins and, perhaps, apes. Broca's area is near the left frontal lobe cortex (or inferior frontal gyrus) and, as just discussed, associated with manual dexterity as well as motor production of language such as speech and sign language.⁸⁵ The second area associated with communication, both verbal and through signing, is located in the left superior temporal gyrus and known as Wernicke's area.⁸⁶ Regarding the hypothesis, also just mentioned, that spoken language evolved from body language, particularly arm movements and grunts that became more complex over time, once again, the magnitude of the capability is a matter of degree. Further, many researchers associate functions in addition to speech with Boca's area since it "supports hierarchical structural processing" and "serves as a functional hub whose capacity for hierarchical computation can be used both for planning sophisticated action selection for manual tasks as well as sophisticated communication."⁸⁷ Thus, overlapping neurological circuits—such as those that support "action selection and the temporal sequencing of elementary actions" that are, in turn "mediated by circuits linking cortical areas (including those in the frontal lobe, such as Broca's) and the thalamus with the basal ganglia"—support language.⁸⁸

Oren Kolodny and Shimon Edelman recently proposed that language evolved through the repurposing of existing neural circuits that supported tool-making in the context of parents teaching children. The specific areas include the cortical (prefrontal and motor) basal ganglia circuits for learning visio-motor sequences in tool manipulation. These same circuits support reinforcement learning. In other words, the plasticity of the brain facilitated the use of the neural circuits involved in tool-making to perform the additional hierarchical task of language development, with the specific activity within the tool-making process that necessitated communication being the instructional phase. This was the ideal situation for language development given the particular plasticity and learning ability of a young brain. The concept of micro-evolution, discussed presently, would have conferred an enormous advantage by facilitating rapid neural and genetic changes in the brain.

Whether language or complex tool-making came first, *archaic Homo sapiens* "had achieved modern or near-modern brain size" by approximately 200,000 ya.[89] This is also about the time that, some scholars and geneticists believe, language began to take root, specifically, with the advent of the FOXP2 gene that is closely associated with advanced language capabilities.[90] Early *Homo sapiens* ancestors did not, of course, wake up one morning and decide to talk to each other. Even with the aid of the "toolkit genes" discussed below (Section 5.8), language had to develop over time along with changes in specific areas of the brain. Given that some of the neurological foundations were already in place that support verbal communication, a relatively long-term force, the nature of which remains a matter of speculation, must have been promoting the repetitive firing of neural connections to establish the needed function.

2.3 Spandrels and Exaptations

In 1979, the renowned naturalist Stephen Jay Gould and his colleague Richard Lewontin introduced the term "spandrel" to describe a phenotypic trait produced as a byproduct of the evolution of a basic characteristic or feature.[91] Originally, the term was used in architecture for the triangular solid space between the top of an arch and supporting beam (known as the architrave). Classical and Renaissance artists utilized or filled in this space with frescos and carvings that enhanced the buildings that they adorned (Figure 15). In 1982, Gould and another colleague, Elizabeth Vrba, proposed the concept "exaptation," which eventually replaced spandrel since it is more consistent with biological terminology generally.[92] In the context of biological (as well as social) development, exaptations are features "that now enhance fitness but were not built by natural selection for their current role." One example, cited by LeDoux, is feathers, which originally developed in cold-blooded reptiles to help control body temperature and later proved beneficial for flight as smaller reptiles acquired this ability.[93] Another example proposed in the literature includes music.[94]

As discussed, Kolodny and Edelman proposed that tool-making, non-verbal communication, and other existing underlying traits provided the foundation for verbal communication, which the plasticity of the brain's neural mechanisms also facilitated.[95] In turn, societal factors, such as the need to plan for hunting, would have reinforced this exaptation. That is, in the context of teaching the art of

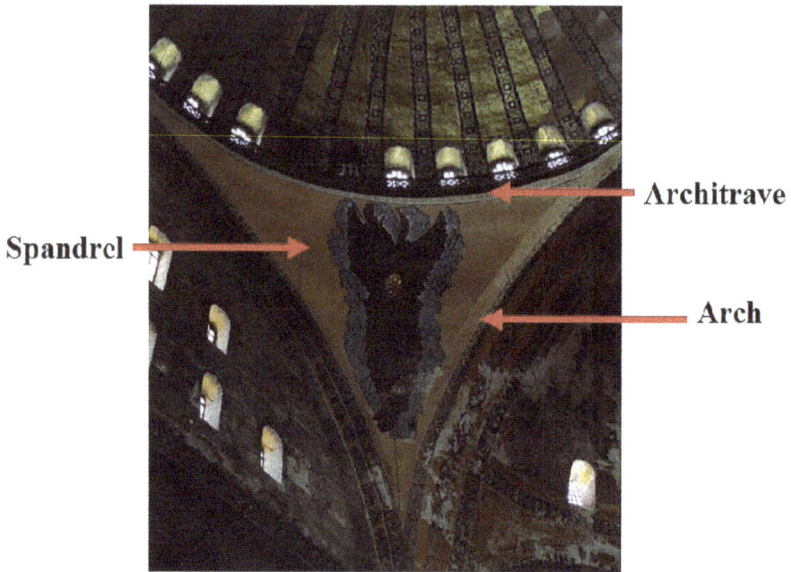

Figure 15. Hagia Sophia in Istanbul, showing the original meaning of "spandrel."
Image credit: photo by the author.

tool-making or knapping, the neural circuits were coopted or adapted to express instructions verbally, while existing circuits that supported meaningful body and arm gestures developed or expanded in their applications to include verbal sounds or phonemes. Since the brains of humans' closest genetic relatives, the great apes, have a Broca's area that supports communication through gestures, the neurological circuit foundations were already in place. So it is that, as noted, deaf people who employ only sign language utilize the same neurological circuits as people who communicate verbally. Likewise, modern sporting activities may also be considered exaptations of features for which natural selection occurred in relation to hunting. That is, the valorization of the precision and accuracy needed to throw with a stone or spear to kill animals that developed in early hominin societies persists in modern society in the entertainment value of spectator sports such as football, for there is little survival value in the skills of throwing, catching, and running today.

The inherent propensity for ritual that humans display can also be categorized as an exaptation in terms of representing an extension of the attribute of pattern completion or patternicity for which the conditions in which modern humans emerged selected. As has been seen, patternicity extends to the establishment of ritual behavior that reinforces, supports, and encourages superstitious and religious belief systems, as will become clear in the following discussion of the notions of cause and effect and belongingness.

3. Cause & Effect

Linked to patternicity is the concept of cause and effect, which can be defined as "the tendency to infuse patterns with meaning, intention, and agency."[96] As noted in the context of Skinner's experiments on pigeons, brains make connections between and find patterns in events and reward themselves for doing so. The pigeons seem to have perceived that their movements prior to pushing the lever and receiving food contributed to the delivery of the reward. Since that was not actually the case, the drive to engage in this behavior can be described as "superstition." Also as noted, humans make similar connections in daily life, knocking on wood or throwing salt over their shoulders in efforts to influence future events and experiencing discomfort when unable to do so at the appropriate moments.

Rituals such as these provide part of the basis for belief systems, including religions. Rituals such as sharing special meals, reciting special words before or after an act, and offering something special to someone or something are fundamental to most, if not all, religions. Rituals reduce dissonance and create meaning for and emotional connections among those who participate in them, helping communities to coalesce. Symbols are among the tools used to reinforce and recreate rituals since they serve to focus and stimulate memories of how rituals are performed and even the belief systems themselves. For example, the menorah and baptismal play a symbolic role in rituals performed in modern synagogues and churches, respectively. Symbols represent a natural step for brains that were already developing language capacity, for language consists of an elaborate system of symbols. Rituals that combine special physical movements, words, offerings, and symbols can create especially powerful emotional experiences.

This propensity to connect events and establish schemas was critical to the survival and evolution of *Homo sapiens*. As previously discussed when our ancestors saw or heard a bush move, it was critical that their first assumption be that it was a dangerous predator. So also for our ancestors, every event or incident has a cause, and something or someone is responsible for initiating any event. Assuming the worst was the safest assumption, and investigating unknowns and finding the answer were critical to survival. Like us, our ancestors felt uneasy when the answer was not forthcoming (that is, the pattern could not be completed). For example, thunder and lightning are unsettling because no explanation is readily apparent for the fact that these phenomena occasionally accompany clouds and rain. This aspect of agenticity is discussed further in the next section. The notion of cause and effect was relevant to tool-making as early humans learned how much force (the cause) was needed to chip a stone to yield the desired flake (the effect). Probably having already learned the amount of force needed to use a hammer-stone to open nuts or the bones of scavenged carcasses to extract the protein-rich marrow, they were prepared to work stone with equal finesse.

As early hominins moved from scavenging to active hunting, their tools evolved into projectiles, especially stones and pointed sticks used as spears. Some primates throw stones, but only hominins have developed the ability to sharpen sticks for use on other animals. Throwing a stone or pointed stick with sufficient force and accuracy to subsist by hunting required practice and expertise developed through trial and error in ways that other animals have never been observed engaging. Further, this expertise has benefits beyond maintaining a supply of food. Even more primitive forms of hunting have profound social effects. Thus, observations of chimpanzees indicate that successful hunters enjoy social influence and, often, preferential access to mating partners.[97] That is, the meat gained from hunting serves as a form of currency that can be exchanged for social and reproductive advantages. The rewards of hunting are also detectable on a neurological level, for hunters experience a surge in testosterone, cortisol[98] and dopamine during the anticipation of a kill. Then, having activated this reward system, the brain generates additional dopamine and serotonin, further reinforcing the drive for success.[99]

Modern humans have retained this drive to hunt and the desire to excel at throwing things with great accuracy that is associated with sexual rewards; as alluded to earlier in the discussion of football, studies have shown that mate selection is influenced by the participation of males in sports, particularly in team sports.[100] The best thrower enjoys great respect within the society (reflected in television interviews and money). Further, football reenacts the tribal teamwork of hunting in the collective effort to achieve a goal and delivers the same hormonal rewards. There are even secondary or tertiary rewards experienced by fans as they root for their teams: male and female fans alike experience increases in testosterone levels, and the fans of winning teams experience a dopamine rush.[101]

Other sports also challenge, exercise, and demonstrate individuals' expertise in judging cause and effect, including archery, firearms target practice, golf, pool, and, of course, hunting. The brain, having evolved this capability to perceive cause and effect, seems driven to exercise it whenever possible. We, as a species, have retained and maintained the drive to understand the proportional interaction between objects. Since these skills, evolved for hunting as opposed to scavenging, are often unnecessary for everyday survival but are left over from a previous phase of our evolution, sporting activities represent a kind of spandrel or exaptation. We continue to engage in them even despite the possibility of injury because of the positive hormonal and cultural feedback with which they are associated.

However, this positive hormonal feedback system may also have negative consequences. For example, gambling addiction has been attributed to increases in dopamine and testosterone levels during the "anticipation" period of the activity, the hormonal activity correlating with the degree of uncertainty in the game.[102] It is the excitement of placing a bet and anticipating a win rather than the actual winnings that keeps gamblers coming back for more. Another phenomenon involving anticipation is in apocalypticism, the belief in some more or less imminent event, such as the arrival of a savior figure, end of the earth, or even overturning of an election. The adherents of these beliefs experience rushes of anticipation and hormonal surges as the predicted time draws near. Once again, individuals whose brains' cognitive executive functions are unable to regulate their hormonal reactions are prone to engage in counterproductive behavior.

3.1 Agenticity

The knowledge of the cause-and-effect relationship helps to explain agenticity, which can be defined as "the tendency to infuse patterns with meaning, intention, and agency."[103] Early hominins learned the relative amount of force needed to pierce the hide of an animal with a spear or bring down a bird with a rock, but they also understood the difference between animate and inanimate objects. Generally, for them, living objects move under their own volition and inanimate objects do not. So, if something moved, either something else must have moved it, or it moved itself and could be alive, having an agent or agenticity. Consequently, every event that included movement must have a cause, and the cause could pose a threat. Once again, the penalty for not perceiving a movement as a threat could be death, while the penalty for hypervigilance in this regard was usually only wasted time and calories. Therefore, most animals, whether hunted or hunting, associate agenticity with movement to decrease their odds of exiting the gene pool early.

Humans have the greatest capacity to assign and explain agenticity. An example that persists into modern times is the calendar. Before television, our ancestors often watched the night sky. As the earth turns, the stars appear to pass overhead in a fixed formation. That is, all of the objects in the sky move together in predictable ways—except, that is, for seven objects: the sun, moon, and five visible planets (Mercury, Venus, Mars, Jupiter, and Saturn). Not coincidentally, there are seven days in the week devised by the Babylonians and adopted by many other cultures including Jews as reflected in the creation story in the book of Genesis.[104] In fact, the number seven has been special or sacred in many cultures and religions. Early humans reasoned that these objects in the night sky had agenticity and were, possibly, alive. Almost all ancient civilizations named and recorded these objects in their documents and told myths about them because their patterns of movement were distinct. The Greeks and Romans are among the many cultures that have deified and prayed to these objects, which became so important that their names are still with us today, on our calendars. Thus, in the English language, three of the days of the week are named after moving celestial objects—Sunday (Sun), Monday (Moon), and Saturday (Saturn)—and the four others are named for various Norse or Germanic gods (Tiw, Woden, Thor, and Freya).

No other animals, not even great apes, investigate moving objects anywhere near as exhaustively as humans do. We call this drive to investigate "curiosity." When an object moves, most mammals in range will notice the movement, continue to watch it for some time, either run away from it or sniff it, and then ignore it if it doesn't occur again. Other animals, like hominins, tend to assume that a thing that moves is alive but, unlike hominins, do not feel compelled to learn what moved it and, as just noted, rapidly lose interest in it if it falls still or avoid it if it smells dangerous. Hominins, by contrast, remain interested until they find an explanation for movement. Otherwise, they are unsettled and plagued by mental dissonance and obsess about the cause of or agent behind movement and, if one is not readily apparent, make one up. This drive to determine or create purposes for objects is referred to as "teleology," which can be defined as the use of design or purpose to explain a phenomenon and, conceptually, includes both agenticity and purpose. Identifying a cause for an action or event relieves hominins of the feeling of dissonance, creating the sense that a potential hazard is at least intelligible, and, at best, not a hazard after all.

A good example of this way of thinking is the aforementioned conceptualization of thunder and lightning. Only in the past few centuries have the forces behind meteorological phenomena—in this case, and the balance of positive and negative ions in the atmosphere—become clear. Electrical storms are a cause for fear as well as a source of curiosity, an example to which I return in Section 4.7.

3.2 Formula for a Conspiracy

The drive to find or invent a cause for everything has several side-effects. Satisfying this drive in effect completes a pattern, an achievement that, as has been seen, the hominin brain rewards with dopamine. Finding a cause, then, answering a question, is not only rewarding but actually addictive. So it is that television programs target viewers with programs that include the word "mystery" in the title and books in the mystery genre sell in the millions every year. So also, conspiracy theories usually involve mysteries and alleged patterns that catch the attention of the public. Simply put, describing a mystery creates anticipation which is addictive, and then offering a possible solution that completes the pattern can be a winning combination—and, therefore, sales—strategy.

It is in this respect that conspiracy theories reflect humans' propensity for patternicity and agenticity. The process of connecting selected pieces of data to form a pattern and attributing it to an agent can mitigate mental dissonance. According to Shermer, the prefrontal cortex seems to exercise executive control over discrimination and decision-making, and the anterior cingulate cortex seems to discriminate among multiple sets of data to choose the best option.[105] The prefrontal and anterior cingulate cortices of a brain that is particularly adept at seeing patterns may be flooded with more options than it can process. As noted, individuals who excel in finding patterns in data tend to have a propensity to attribute agency to events as well as for spirituality and religiosity. Consequently, those who find purpose and agenticity in patterns may look to a spirit or deity as a cause. Notably, the susceptibility to conspiratorial thinking often correlates with the feeling of lacking control over external events, for creating an agent and purpose behind what seems like chaos serves to give events some sort of order and meaning.[106] Again, the hormonal reward system can, if insufficiently regulated, become overactive and create false belief systems with potentially dangerous results.

3.2.1 Conspiracies and Conspiracy Theories

In making arguments such as these, it is important to distinguish between conspiracies and conspiracy theories. Actual conspiracies occur that can be documented by the confessions of those who have participated in them and other forms of proof. Conspiracy theories, on the other hand, are usually less well-documented, wider in scope with regard to the number of participants, and lack documentation regarding both the identities of the participants and the connections among the various pieces of data that participants adduce in support of their conspiratorial beliefs..

I suggest, therefore, that conspiracy theories develop

primarily as a result of the confluence of three proclivities that have evolved in *Homo sapiens*. First, as Shermer argued, humans' "pattern recognition filters" are wide open.[107] When a horrendous event occurs, many who experience it are left feeling that their lives are out of control. This loss of control correlates with the perception of being in danger and initiates the drive to identify the nature and source of the threat. Consequently, the pattern recognition system goes into overdrive, searching for more data points than the senses have provided. Every minute detail in the context of the precipitating event is examined to determine, not whether it fits in, but how. The neurological reward for completing the pattern of the conspiracy is, again, dopamine, delivered by the brain in levels proportional to the complexity of the pattern.

Second, humans' exaggerated tendency toward agenticity reinforces the teleological perspective, the need to find causes commensurate with major events.[108] So, when President Kennedy was assassinated, the need was felt for a cause or reason beyond simply a single assassin. When the World Trade Center was destroyed, assigning the responsibility to just a few terrorists was unsatisfying, so many looked to the government. Under such circumstances, the proclivity to find agenticity kicks into high gear and generates dissonance when an agent cannot be identified. So it is that most belief systems employ teleological arguments, reasoning based on the concept described above that all events have a purpose and that they are leading to an intended end in the fulfillment of a grand plan that may not be apparent but is discoverable or will be revealed. This reasoning applies not only to conspiracies but to apparent accidents. When an airplane crashes, many believe that it was "god's will" above and beyond any mechanical failure. In religious environments, believers have faith that there is a purpose to everything, just as, in the QAnon environment, believers "trust the plan."

Third, human thinking is subject to confirmation bias, that is, the tendencies to seek and accept data that supports preexisting ideas.[109] Thus, people tend to be receptive to data that support their established emotional schemas. Those with negative preconceptions or emotional schemas relating to the government may be inclined to accept data that, for instance, seem to support the theory that the Bush administration was complicit in the terrorist attacks on September 11, 2001, or that Vice President Johnson had a hand in the assassination of President Kennedy.

Conspiracy theories often share general themes. To begin with, they tend to involve the machinations of a secretive, well-organized, and powerful organization in pursuit of some goal. Organizations that are already secretive, such as the CIA and the Mafia, for example, therefore play key roles in many conspiracies. Another theme is the involvement of wealthy individuals, such as the businessman George Soros, or groups, such as the Bill Gates Foundation founded by the software magnate, the reasoning being apparently that great wealth brings power and the desire to pull the strings behind the scenes. The government bureaucracy that persists through changes in administrations, sometimes derided as the "Deep State," is also a natural candidate for a role in conspiracy theories. Of course, the ultimate conspirators are the gods, who tend to be represented as powerful, invisible, secretive, and acting in fulfillment of some overriding plan.

3.3 Modern Homo sapiens

Something changed in the brains of some hominins approximately 150,000 ya. The change was not abrupt and is perhaps better described as a transition. The cause may remain unknown, but significant behavioral changes begin to appear in the archaeological record, including the use of much more diverse tools for hunting, self-adornment (indicating an increase in self-awareness), and sophisticated burials (including special objects and personal items). Notable in terms of social organization, among the interred are disabled individuals who clearly had received long-term care from, presumably, members of their families and tribes. These behavioral changes indicate changes to the brain that neurologists and anthropologists have been able to describe.

Around 60,000 ya, groups of modern *Homo sapiens* began to migrate out of Africa to people, eventually, all of the continents except Antarctica.[110] *Homo erectus* migrated out of Africa earlier and *Neanderthals* had populated much of the middle east and southern Europe but both were ultimately replaced by modern *Homo sapiens*. These more recent hominins had the ability to think symbolically, which they expressed in various forms of art, including paintings of figures on cave walls and carvings of three-dimensional figurines.

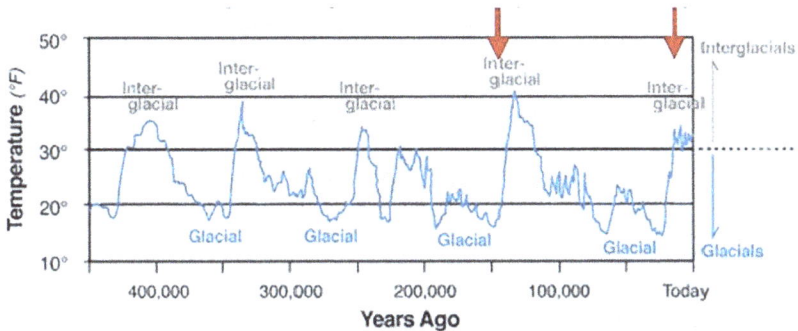

Figure 16. Glacial-interglacial cycles over the past 450,000 years. *Image credit: courtesy of the Utah Geological Survey.* [iii]

The climatological record shown in Figure 16 may offer some clues regarding the motivation for these changes. Notably, an interglacial warming period began approximately 150,000 ya. The retreat of the glaciers increased the amount of open land and the availability of flora and fauna, expanding the hunting opportunities for hominins. These conditions brought, in turn, increased migration opportunities as well as increases in hominin populations. Though this climate change correlates with the behavioral changes in *Homo sapiens,* there is no absolute proof of causation. I will return to this discussion in the context of the next and most recent interglacial warming period, which began approximately 20,000 ya, presently.

4. Self, Empathy, and the Theory of Mind

Over the period from 150,000 to 100,000 ya, the brain of *Homo sapiens* underwent profound internal changes. These changes were associated with behavioral changes that are reflected in recently discovered archaeological artifacts. Over this period, people began to make and use a much more diverse array of stone tools than before. They also began to acquire a deeper awareness of the self than *Homo erectus* and early *Homo sapiens* that made them aware of and concerned about others' opinions of them. They began to develop a much greater sense of empathy and Theory of Mind (again, ToM). While it remains unknown what precipitated this transformation, the corresponding changes in the brain necessary for these behavioral developments are becoming clear.

4.1 The Self

As discussed in the context of the brain's ability to create mental models of objects, pieces of information relating to a given subject or event are bundled together in the mental models or patterns that neurologists and psychologists call schemas.[112] Schemas also serve to generate expectations about objects, events, and situations, making them the foundation or building blocks of cognition.[113]

Also as discussed, many scholars associate tool-making with cognition. In fact, many argue that cognition is necessary for tool-making. Since the ability to create and recall schemas in the mental working memory is the foundation for cognition or consciousness, and since other species have limited tool-making capabilities, it follows that they have some cognition or consciousness. This is difficult to determine given our limited ability to communicate with them. It is certainly the case that the brains of humans differ from those of other species with respect to both internal organization and cellular function. Once more, the presence or absence of consciousness is probably not binary (i.e., an organism either has it or not) but better described as a gradient or continuum.

It is useful to review briefly here some of the key points relating to brain structure (Section 1.1), beginning with LeDoux's comparison of the human brain with the brains of apes. First, the human prefrontal cortex shows distinct patterns of connectivity "within and between cell layers." Second, the interconnectivity between the neurons in the prefrontal cortex and those in the parietal and temporal cortices is "stronger" in humans than in apes. Third, gene expression in the human prefrontal cortex shows distinct patterns of "energy metabolism and synapse formation" not observed in apes.[114] Fourth, the granular cells found in the fourth layer of cells in the paleocortical tissue of primates, which humans use for working memory and cognition, are found in "the [human] dorsal and ventral lateral prefrontal cortex and the frontal pole."[115] Torrey pointed also to VENs (von Economo neurons), also known as spindle neurons, which "are thought to be a phylogenetically recent specialization in human evolution" and "have been called 'the neurons that make us human.'"[116] Found primarily in the anterior cingulate and anterior insular, VENs offer supporting evidence that self-awareness is a matter of degree among mammals, for they are found to a lesser degree in the brains of the great apes as well as elephants and dolphins, all of which have shown some degree of self-awareness in mirror tests. Lastly, human brains feature the frontal pole structure in the prefrontal cortex that may be a primary attribute of the enhanced consciousness necessary to develop tool-making—or, alternatively, tool-making encouraged the development of the frontal pole structure. In any case, as part of the prefrontal higher-order network, this structure is "connected with the dorsal and ventral lateral prefrontal areas, as well as with the multimodal convergence zones in the neocortical parietal and temporal lobes, and with medial temporal lobe areas."[117]

Thus, early hominins, including *Homo habilis* and *Homo erectus*, seem to have had a sufficient sense of self-awareness to produce tools. Along the continuum of such capabilities, these species and early or archaic *Homo sapiens* had a less-developed sense of self-awareness than modern humans but a greater degree of self-awareness than earlier hominins and the great apes. These considerations are relevant here because the "self-schema" is among the crucial schema that the human brain creates, and the frontal pole has "been implicated in perceptual self-awareness and the enhanced ability to use autobiographical memory in thinking about one's self."[118] The frontal pole has access to the state of one's body and schemas representing it as well as the brain itself. It is, in other words, the executor of self-process that utilizes the autobiographical memories and data coming from the hippocampus and other areas in the brain, as shown in Figure 17.

However, personal representations of one's self or self-

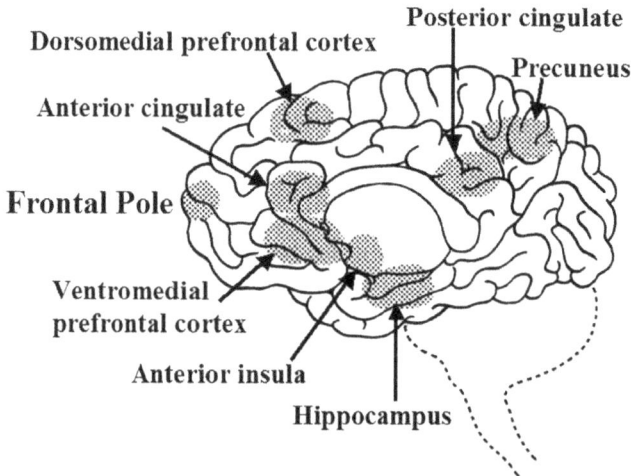

Figure 17. The frontal pole and other structures relating to the self. *Image credit: drawing by Randy Mohr.*

image as an individual seem to reside primarily in the five parts of the brain, all of which the frontal pole accesses, shown in Figure 18. These areas are the dorsal medial prefrontal (involving the executive self and self-reflection and mentalizing), anterior cingulate (the executive self), ventromedial (feelings of ownership), and orbital frontal cortices and the anterior insula (body states).[119]

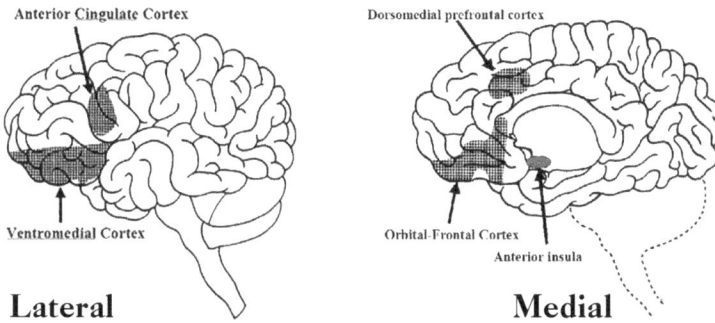

Figure 18. Parts of the brain associated with the self,
Image credit: drawing by Randy Mohr.

As discussed, Homo sapiens' sense of self-awareness developed from about 150,000-100,000 ya, as articles of self-adornment or personal identification found in the archaeological record demonstrate. An example is a set of 33 shell beads from Bizmoune Cave in Southwest Morocco. Many of the beads come from deposits dating to approximately 140,000 ya making them the oldest shell beads yet recovered. This and similar atifacts document an enormous leap in capability, indicating that these hominins were, as described earlier, not only aware of themselves and others but also concerned about others' opinions of them. Through self-adornment, individuals attempted to communicate to others something about themselves that marked them as special or significant with respect to clan identity, physical attractiveness (facilitating reproduction), hunting prowess, status, wealth, and so on. The concept of clan identity includes belongingness, which is an extension and major part of individuals' sense of self that also emphasizes distinction from some "other." To the category of "other" belong individuals outside the clan, identifiable as those who do not wear similar adornment. In other words, for the humans of 150,00-100,000 ya, symbols such as the Bizmoune Cave beads served to establish a sense of self as well as belonging to a group, the dimensions of which could even transcend traditional hunter-gather bands and extended family clans. The awareness of the self as both an object and the central subject of any activity is essential for the concept of the mind if not the mind itself. From this perspective, the conscious awareness of the self and self-image may be thought of as "the mind," a term long in use and, therefore, burdened with considerable baggage. The imagination or consciousness can have thoughts and dreams independent of real-time sensory data (sights, sounds, etc.) and set them in the near or distant future in great detail. For these reasons, the mind has often been considered separate from the body. Philosophers and theologians have long entertained this notion of dualism. So-called "monists," by contrast, consider the mind inseparable from and, indeed, of one and the same substance as the brain's neurological processes.[122]

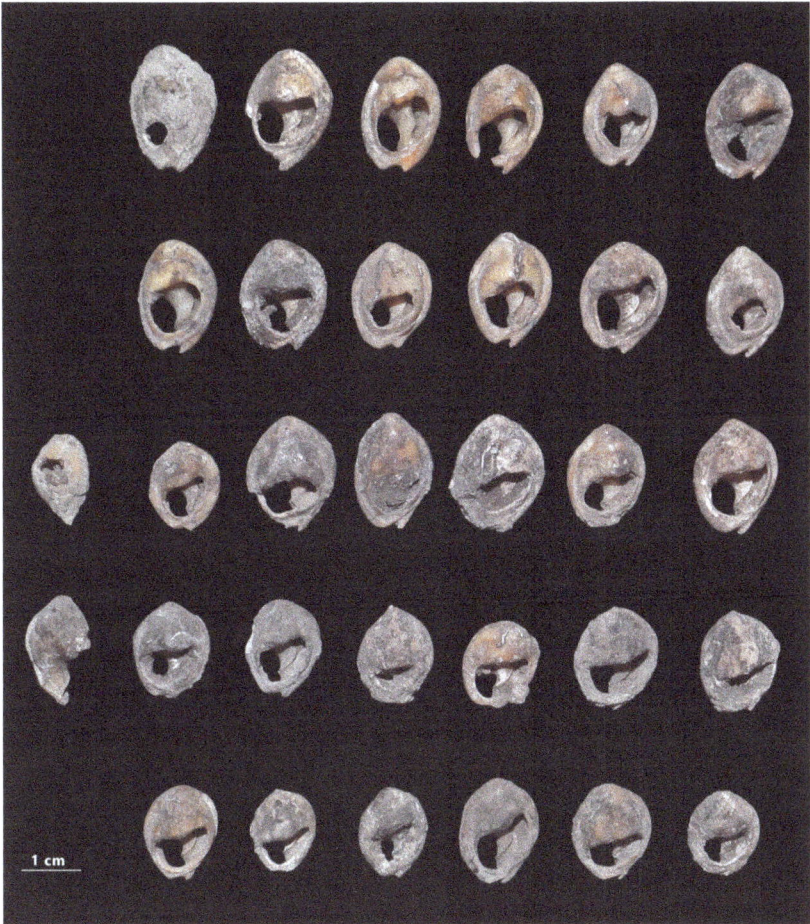

Figure 19. Personal Ornaments from Bizmoune Cave, Essaouira, Morocco, 140,000 ya.
Image Credit: A. Bouzouggar, INSAP, Morocco. Used with permission.

4.2 Emotions

Closely related to the self are the emotions. Most neurologists agree that there are no emotions without the

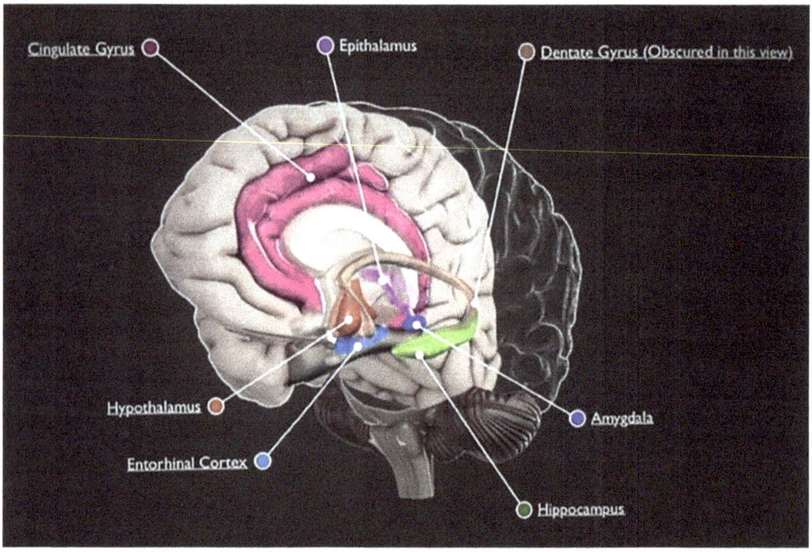

Figure 20. The limbic system.
Image credit: Cold Spring Harbor Laboratory DNA Learning Center 3D Brain app.

awareness of 'Self.' Lower animals react to things or situations, but their range of responses is limited to such basic emotional states as fear. As previously stated, memory resides in large part within or near the limbic system, which has also long been identified as the center of emotions. The hippocampus and amygdala, which are critical for the formation and storage of memories, are also part of the limbic system (Figure 20).

Combining the above observations, it is apparent that seven areas of the limbic system contribute to humans' perceptions of the world and, thus, help to determine human behavior: the hippocampus, amygdala, ventral tegmental area (VTA), nucleus accumbens septi, hypothalamus, orbitofrontal cortex, and DLPFC.[123] The hippocampus, amygdala, and orbitofrontal cortex are of particular relevance to the present discussion. The hippocampus, as mentioned, serves as the primary memory bank, and one of the primary functions of the amygdala is "to process sensory information in terms of its emotional significance for the organism."[124]   The DLPFC is critical as the executive function in terms of prioritizing behavior and adapting to change, internalizing social mores throughout childhood, and then serving as the regulatory authority or mediator of behavior such that it conforms to the prevailing traditions and laws.[125]   On a personal level, individuals tend to feel strongly about moral beliefs since the emotions are closely tied to opinions about issues such as, to take a random example, child abuse and, more generally, religious and political matters. These values, rather than being inherited, are inculcated in individuals by their families and the cultures in which they grow up.

The brain, including the limbic system responsible for an individual's "mood" or emotional state, continually monitors the state of the body. Many feedback or loop circuits exist within the limbic system. Especially pertinent to this discussion is the system for monitoring mood and emotion, for it impacts memories in two significant ways. First, as discussed, when sensing something for the first time (for instance, visually), the brain captures the event, including the external context, and associates with it the emotional state at that moment as a new episodic memory for storage in the hippocampus. This complex is precisely what LeDoux means by an emotional "schema."[126] When the initial event, object, or person is frightening, this trauma becomes part of the schema.

Second, when encountering the same event, object, or person again, the "re-cognition" includes the initial emotional state, in this case, fear. Thus, people who are bitten by the first dog that they encounter may fear all dogs, or even four-legged animals generally. Eventually, the executive function in the DLPFC usually adapts to the reality that neither four-legged animals generally nor dogs in particular are likely to cause harm. This process indicates the operation of a learning feedback loop involving the hippocampus, amygdala, and DLPFC.

Many memories, then, have emotions attached to them. Episodic memories especially tend to associate significant objects, people, and contexts with emotions. Noteworthy here is a neural circuit, called the Papez circuit, that runs from the hippocampus and anterior nuclei in the thalamus to the cingulate gyrus in the cortex and back to the hippocampus. This feedback loop facilitates moment-to-moment memory and emotional updating of experiences.[127] As noted, an episode is initially stored in the hippocampus and amygdala in the limbic system. Over time, an episodic memory can transform into a semantic memory. In this process, the objects seem to be transferred to areas of the cortex, often with the loss of the associated emotions. Alternatively, the memory may already reside in the cortex but then diminish in the hippocampus and amygdala, again with the attenuation or loss of the associated emotion.

The connection between memories and emotions influences the former as well as judgments and decisions to which given memories are relevant. In fact, Antonio Damasio found that the ventromedial prefrontal cortex is necessary for "both reasoning/decision making, and emotion/feeling, especially in the personal and social domain,"[128] which is why damage to this part of the brain can result in the inability to make decisions and feel emotions. As observed in the previous section, dualism is the concept that the mind is separate from the body. A related concept in Western philosophy and psychology, discussed since at least the time of Descartes,[129] is that "emotions and reasoning" are "very separate."[130] Descartes argued that behavior is the product of the body under the influence of morals, which he considered to be the product, in turn, of reason and to reside in the mind. Research over the past 30 years has, then, shown to the contrary that emotions are not simply closely tied to but necessary for judgment and decision-making.

The clinical evidence for this conclusion includes a renowned case of a clinical patient of Damasio named Eliot. Surgeons removed much of Eliot's prefrontal cortex to treat a brain tumor, after which he lost the ability to feel emotion, though his IQ remained unchanged and he retained memories of emotions felt prior to the surgery. A further consequence of the surgery was that "Eliot could not make rational decisions any longer. He could discuss the pros and cons of any options presented to him" but "could not weigh the different options. He had nothing that guided him to choose one option over another."[132]

According to LeDoux, initial episodic memories also include the self and may be associated with feelings of ownership that are formed along with the schema of any event that initially forms a memory.[133] Consequently, a fear schema may instigate the flight-or-fight response to things or situations, and these "feelings" attached to a schema are instrumental in the formation of attitudes and belief systems. Once an initial emotional schema is attached to a situation, object, or person, it is very difficult to alter or remove it. Post-traumatic stress disorder (PTSD) is a well-known example of the difficulty of changing emotion schemas that are attached to events.[134] In the words of LeDoux, "emotion and self-schema overlap considerably. Your fear schema, for example, is yours, having been sculpted by things you have experienced and stored in memories. No one else experiences fear exactly the way you do."[135] Consequently, the presentation of facts to counter a person's fear is usually ineffective unless the facts happen to be relevant to the specific situation in which the fear schema formed. For those suffering from PTSD, the simple fact that the situations that generated their trauma no longer exist does not mitigate either the memory or the associated fear. Likewise, since profound spiritual experiences or episodes are often associated with positive emotions, the inconsistency of any such experiences or episodes with the laws of physics is not relevant to the perceptions and beliefs with which they are associated.

Another crucial executive function of the limbic system with regard to decision-making relates to its interactions with the executive functions in the prefrontal cortex and other cortical areas. These interactions mean that, again, most, if not all, decisions involve emotions. Patients with injuries to the limbic system where the emotional schema reside have trouble with decisions, which require the executive function provided by the prefrontal and anterior cingulate cortex.[136] These observations raise the question of whether either free will or dispassionate life-changing decisions are possible.

4.3 The Other

As has been seen, human evolution has involved distinguishing animate and inanimate objects based on the possession or not, respectively, of a life force with potentially nefarious intentions, agendas, or agenticity. Relatively recently, hominins developed the concept of identity or self and, with it, the understanding that others of their kind and even animals may also have a self with similar but distinct agenticity, which is to say, a Theory of Mind (ToM).

An important parallel or precursor for the development of a ToM is the capacity for empathy, which is facilitated in the brain by the presence of mirror neurons and further reinforced through socialization. Mirror neurons are distributed throughout the brain, including in the motor cortex,[137] and are responsible for, for instance, the contagiousness of a yawn in a group of people or flinching when observing another individual being struck. These types of neurons are also distributed in the limbic system, again, allowing for the perception of emotions in others and feelings of empathy. Mirror neurons are present in the amygdala as well; in fact recent studies show "that individuals with smaller amygdala are more likely to present as having sociopathy."[138] Other animals are also known to display empathy and compassion;

what is distinctly human, as with some other human attributes, is the magnitude of the experience of empathy of which humans are capable. For example, only rarely in the wild do healthy chimpanzees take care of disabled chimpanzees (apart from mothers and their young),[139] though there are recent reports of bonobos adopting infants from a different troop.[140] Hominins, on the other hand, routinely take care of disabled members of their group or tribe, even for relatively long periods of time. This behavior is, though, relatively recent, having, as with certain other advanced or modern capabilities of *Homo sapiens,* come to be evident in the archaeological record after about 80,000 ya.[141]

Fossil and skeletal remains offer evidence for an increase in hominins' capacity for empathy. Thus, quite a few examples of human long bones have been found that show fractures that had healed and continued use of the limb for many years thereafter. A well-known example is a hominin skeleton from a cave in Shanidar, Iraq excavated in 1957. (Figure 21) A recent reevaluation of these bones by Trinkaus and Villotte indicates that one humerus was fractured at least twice during adolescence and underwent a non-union fracture and possible amputation at some point but that the individual survived into his 30s or 40s.[142] Skeletal remains from caves in Spain and Iraq dating from 60,000 to 80,000 ya include other individuals with multiple fractures that had healed and some who suffered from severe arthritis.[143]

The implications of these finds are enormous. The individuals with the healed fractures had suffered injuries that previously would have been lethal but had survived to hunt again after recovery that would have involved long-term care. Thus, they must have been taken to some sort of shelter, protected from predators, and fed for several months while they healed. Those with crippling arthritis must have resided at semi-permanent campsites where they received protection and, presumably, benefited from a division of

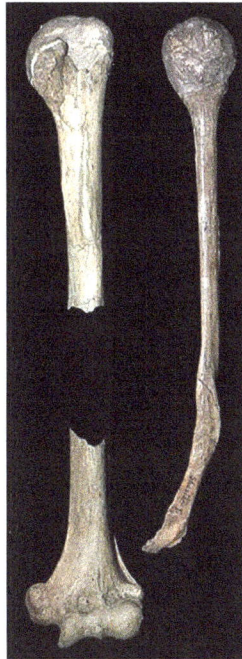

Figure 21. Left and right humerus from skeletal remains
of a hominin found at Shanidar, Iraq, 60-45,000 ya.
Image credit: photo by of Erik Trinkaus
(used with permission and cropped by the author).

labor that allowed them to remain valuable to the group despite their condition. All of this evidence indicates a giant step toward complex social order, as discussed in the next chapter (Section 5.3).

Mirror neurons also facilitate the sense or perception of "the other,"[144] without which humans would "probably be oblivious to the thoughts, emotions, and actions of other people."[145] This empathetic perception that other individuals have thoughts, pain, feelings, and intentions facilitated the giant step in respect to the ToM. To review, the ToM, as a philosophical concept, traces back at least to Descartes. To a considerable extent, it forms the basis of the theory that the mind is, or can be, separate from the body, as Descartes maintained. It seems fairly clear how this concept developed naturally. Neurologists are beginning to suggest that dreaming plays an important role in learning and retaining information and events.[146] As noted earlier, dreaming may be a useful or necessary function for many species possessing a cortex. In humans, it may also contribute to the ToM, since, for example, it is possible to dream of being in another place, even far away from where the sleeping body resides. As a personal example, my partner told me one morning that she had dreamed of being with her grandson, during which time she was physically distant from him while, in her mind, she was at his side. Similarly, when loved ones die, it may be comforting to believe that their minds have gone somewhere else.

The ToM, then encompasses the understanding that others have minds with understandings and motivations that may differ from one's own. A ToM allows people to attribute thoughts, desires, and intentions to other people, animals, and even inanimate objects, to predict or explain their actions, and to posit their intentions—in other words, to understand that mental states can be the cause of, and, thus, account for, the behavior of others. Animals learn very early the difference between animate and inanimate objects, and hominins soon learn that moving objects may have intentions or motivations that differ from their own. The perception that anything that moves has energy and, possibly, a life of its own, and can be friendly or dangerous, underlies the basic fight-or-flight response to sudden changes in the environment. In the early development of humans, as social animals, it was a matter of life and death to determine whether another was friend or foe very quickly.

An even higher level of the ToM is the understanding that others have a ToM as well as this same understanding: you know that they have it, and they know that you know that they have it. The realization or assumption that other creatures have a mind is obviously distinct from the basic knowledge of the self. Like the self, though, the ToM resides in

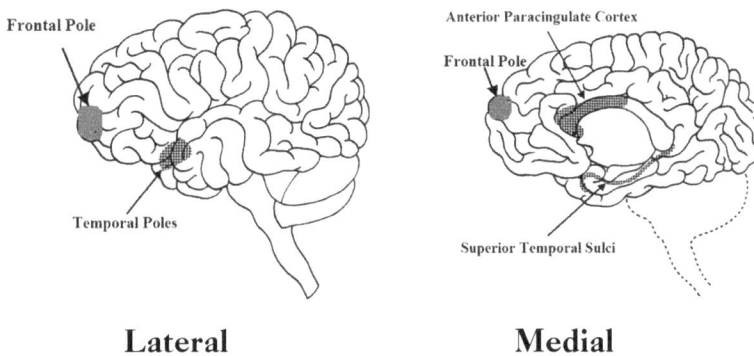

Lateral   **Medial**

Figure 22. Areas of the brain where the ToM resides.
Image credit: drawing by Randy Mohr.

several areas of the brain:[147] the anterior paracingulate cortex, the superior temporal sulci, the temporal poles on each side of the brain, and the frontal pole (Figure 22).

4.4 Anthropomorphization

Since people only have their own minds as a reference, they have a propensity to project human characteristics not only onto other people but also onto animals and even inanimate objects that move, such as planets. This is the phenomenon of anthropomorphization (from the Greek ανθρωπος, "man "[148]). Most pet owners project human thoughts and emotions onto their animals and, consequently, human intentions and motives and even the capacity to understand complex concepts. The difficulty in learning what a pet is actually thinking is its lack of language. People in antiquity experienced the same challenge regarding inanimate objects such as the planets: unable to communicate with them, they assumed that certain objects had intentions, agendas, and emotions comparable to their own. Indeed, many believed that to assume otherwise was dangerous. Consequently, they wanted and needed to keep the gods happy so as to avoid their wrath, as the mythologies of many cultures demonstrate. These myths are best known to contemporary Westerners in the context of the stories and sacred buildings created by the Greeks, Romans, and Jews. Their temples often featured altars nearby where sacrifices were burnt because the gods were thought to enjoy the aromatic smoke.[149]

4.5 Disembodied Minds

If the mind were separable from the body as 'Dualists such as Descartes' believe, then it is not hard to conceptualize minds that don't have bodies which seems to be confirmed in dreams. It is possible to interpret the images that appear in dreams as spirits, perhaps of ancestors, that may have friendly or unfriendly intentions or motivations.

The ToM begins to develop very early in childhood and is nearly complete by age four.[150] Babies learn the difference between animate and inanimate objects and between animate objects that are biologically alive and those that move by mechanical force. Children also begin to grasp that the mind is separate from the body and to entertain the notion that minds can exist without bodies. This concept is fundamental to play, specifically, pretending.

Around the turn of the century, Marjorie Taylor interviewed a group of children and their parents from a religious background their opinions regarding imaginary friends and mythological figures such as Santa Claus. As Figure 23 shows, 85% of the children whom she interviewed under the age of 5 said that they believed in Santa Claus and 90% that they believed in god, while 28% said that they believed in an imaginary companion. Of those aged 5 through 7, 65% believed in Santa Claus, 90% believed in god, and 63% also said that they had had an imaginary friend at some point. However, none of those over the age of 7 admitted to having an imaginary friend, and only 25% professed belief in Santa Claus, though, again, the belief in god remained constant. After the age of 12, the belief in Santa Claus also dropped to virtually zero; once more, the majority continued to profess belief in god. The steady decrease in the belief in Santa Claus over time may have been due to information provided by parents or siblings or the application of logic. In many families, belief in god is sustained or reinforced by parents and the community while Santa Claus is, for those over the age of 12, primarily associated with commerce.[151]

Age	Santa Claus	Imaginary Companion	God
Birth-4	85%	28%	90%+[152]
5-7 years	65%	63%	90%+
8-12 years	25%		90%+
13-Adult		*	90%+

*An adult with an imaginary friend is diagnosable as a paranoid schizophrenic (unless the imaginary friend is god).

Figure 23. Changes in beliefs about supernatural figures with age.

4.6 Dreams

Scholars largely agree that dreaming contributes to the belief that the mind is separate from the body. Most contemporary hunting and gathering societies, as well as many modern western cultures and religions, have the further belief that the mind, or soul, leaves the body at death and goes to some type of spirit world. Two types of dreams reinforce these beliefs, visitation by deceased family members or leaders and visits to distant places, with distant people, or with spirits.[153] Additionally, visions and hallucinations induced by compounds in mushrooms and other bioactive substances can reinforce the belief in mind-body dualism.

Dreaming, as noted, is not limited to *Homo sapiens*.[154] Other species exhibit dream motor functions such as rapid eye movement (REM) and the movement of feet or paws. REM sleep does seem to be exclusive to placental and marsupial mammals.[155] The purpose of dreaming is hotly debated, and there are many theories. Michael S. Franklin and Michael J. Zyphur argued that dreams serve as a virtual rehearsal mechanism, helping dreamers to handle situations and encounters successfully. Regarding the significance of dreams for the belief in mind-body dualism, Torrey stated that: "hominins could not assign meaning to their dreams until they had cognitively matured. Specifically, they needed to have acquired an awareness of self, awareness of others, introspection, and the ability to place the experience of their dream within the context of their past experiences and future hope."[156]

However, in order to confirm these experiences through sharing with other hominins and, ultimately, develop an ancestral or spiritual belief system, language is necessary—a topic to which I return in the next chapter (Section 5.1.1-2).

4.7 Formula for a Deity (or Deities)

Human children, at least in the western tradition, then, come to believe that the mind is separate from the body or brain, as demonstrated by their belief in imaginary friends, and dreaming reinforces this belief as they grow into adulthood. Today, neurologists and psychologists are documenting the selective evolution of the brain to experience specific tendencies, proclivities, and propensities that facilitate the belief in a deity or deities.[157] These include, as discussed, the desire or drive to complete patterns, which extends to ritual activities, the assumption that every event has a cause, the teleological perspective that everything has a purpose and function, and the anthropomorphization of virtually anything that moves. Disembodied minds can be more powerful than those encased in flesh since they can outpace physical bodies and may be immortal. However, seeing these entities only in dreams, and unable to interview or question them while awake, humans are left to imagine for themselves the scope and magnitude of their power. We further tend to believe that events and even objects have a purpose or function. As previously mentioned, using design or purpose as an explanation of natural phenomena is termed 'teleological.' Regarding specifically teleological thinking, studies of children reveal that they generally view natural phenomena as intentionally designed by a god. Thus, Deborah Kelemen reflected,

Not coincidentally, they therefore view natural objects

as existing for a purpose . . . . [The] current data on children's promiscuous teleology and explanations of origins might therefore suggest an obvious affirmative answer to the question of whether children are intuitive theists.[158]

Returning to the discussion of cause and effect (see Section 3.1), *Homo sapiens* have evolved the neurological drive to complete patterns and determine causes discussed above. Regarding the example of electrical storms, early hominins would have had a pretty good idea of the force necessary to knock down a tree but difficulty identifying the cause behind the force. They probably reasoned that it must have been alive since inanimate objects don't seem to behave in this way—so, it was a disembodied mind. Since they could not see or understand what caused lightning strikes and the associated thunder, they deduced or invented a cause. They were probably also concerned not to offend something so powerful. This disembodied force was, presumably, male since males in their experience tended to be more powerful than females. In other words, there existed an invisible, disembodied mind in the sky that occasionally sent down a fiery force powerful enough to destroy trees. Numerous ancient gods were credited with such events, including the Greek Zeus and the Hebrew YHWH.[159]

Given the need to provide causes for events, the belief in the possibility of disembodied minds, the belief that something or someone must cause natural phenomena such as lightning and thunder, the human tendency to anthropomorphize and believe in disembodied minds, and the understanding of the enormous amount of force that natural phenomena can unleash, it is not difficult to understand why our ancient ancestors came up with the idea of Zeus the Thunderer. Zeus was often depicted holding a lighting bolt. We have evolved an extremely fertile imagination that has not only ensured our survival and facilitated our emergence as the apex predator in the animal kingdom but also made us spiritual, superstitious, religious, and capable of developing elaborate conspiracy theories.

4.8 Where the Gods Live

As discussed, fMRI technology makes it possible to see which areas of the brain are active when someone thinks about what someone else is thinking (Figure 24). We might then wonder whether there are areas of the brain that are active when someone is thinking about (a) god.[160] A 2009 study by Kapogiannis et al. found this to be the case. Hence, in the words of the title of their article, to the brain, "God Is Just Another Guy." These researchers reported that: "Brain scans of participants thinking about god show activation in the same part of the brain where people empathize with others. One such region, called the precuneus (the upper green dot on the scan), is also associated with imagination, balancing complex tasks and self-consciousness."[161]

This finding makes sense since a human being has no more idea how a divine brain or mind works than that of an animal. Or, rather, people have a slightly better sense of what an animal's thoughts might be based on comparative analyses

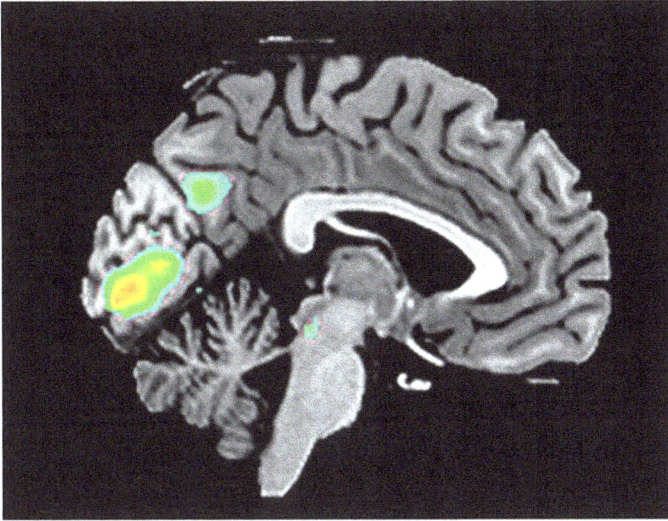

Figure 24. fMRI of the brain of a person thinking about another person's thoughts.
Image credit: photo courtesy of Dr. Jordan Grafman.

of the structure of animal and human brains, and numerous behavioral tests for animals have provided a glimpse of their mental capabilities. To date, it has not been possible to submit the brain of a god to such analyses and tests to determine its potential. Consequently, people anthropomorphize a god's thoughts by projecting their own. This is the primary reason that holy scriptures often depict gods with human emotions and as dispensing rules for human actions and even thoughts.

The science shows, then, that, when people think about what their god or gods are thinking, the latter reside in the precuneus and the parts of the brain that are active in relation to the ToM.[162] Not all people have a god or gods residing in their brains, but, for those who do, the place of residence is here. However, conjuring up the image or images of a god or gods means uploading images from memory into the mental workspace, as discussed, including the occipital lobe, which is the seat of the imagination. The sources of such images usually depend on the particular religious training to which individuals are exposed, often in childhood. Many concepts of the embodied physical god derive from images that children see in places of worship, for first impressions are often the strongest.

Michael Gazzaniga has also identified the temporal lobes, particularly the left one, as the primary locus of religious activity. This, in his words, "left-hemisphere interpreter" generates stories about life. Humans seem to have a predisposition to fit new information and events into preestablished stories, or their "worldview," as I understand the concept. Gazzaniga offers clinical evidence for the significance of this area of the brain in the form of temporal lobe epilepsy (TLE). The symptoms of this type of epilepsy may not manifest externally apart from some repetitive behavior or the appearance of being dazed. Notably, the behaviors that the neurologist Norman Geschwind identified as consistent with TLE include: "(1) hypergraphia, the tendency to write compulsively and copiously; 2) hyperreligiousity, the tendency to be extremely religious, with a great concern for morality . . . 3) aggression, usually transient and not leading to violence; 4) stickiness, or dependence on others, or clinginess—for example, being unable to end a conversation; and 5) altered sexuality, either increased or decreased, but to an extreme."

Geschwind further diagnosed several famous historical figures, including Van Gogh and Joan of Arc, as TLE sufferers.[163]

People and societies like to share their gods. In fact, many scholars have suggested that one of the purposes of religion is to facilitate the coalescing of groups from various tribes characterized by distinct customs under common laws, morals, and customs. Thus, Kapogiannis et al. concluded that "religious belief uses a brain system that evolved quite recently"—perhaps as humans evolved the ability to handle complicated social interactions over the past 60,000 years.[164]

One of the best ways to share or display divinities is through imagery, whether two- or three-dimensional. Early three-dimensional motifs or icons—those made before approximately 6,000 ya—were most often small female figurines. It is unclear whether these figures represented gods, but they do seem related to fertility. Since 6,000 ya, many Western religions have portrayed the primary god as an old man brandishing symbols of power such as a lightning bolt or hammer. These gods are represented as old because they were conceived as having created the world in the distant past.

4.9 How Are These Beliefs Perpetuated?

Religions are persistent and pervasive, and they can confer evolutionarily significant social benefits. Otherwise, there would have been selection against religion over time. In other words, something that is as elaborate, institutionalized, time-consuming, energy-consuming, resource-consuming, thought-consuming, and a frequent source of violent conflict would not exist in the absence of the capacity to provide some "secular value."[160]

Toby Lester has studied religions in an effort to determine why some of them prove successful and others die out. He observed that those that persist and grow "promote health, mate selection, and security," emphasizing "caring" and "helping people survive." For example, most of the main religions today emphasize charity and have some form of the "Golden Rule," often expressed as "do unto others as you would have them do unto you."[166]

Beliefs in specific powerful disembodied minds, complete with names, are passed down from generation to generation. One way to perpetuate a belief is to tie it to the limbic system, thereby ensuring that emotions sustain it, sometimes even in the face of overwhelming contradictory facts. Thus, Shermer stated that: "Once beliefs are formed, the brain begins to look for and find confirmatory evidence in support of these beliefs, which add an emotional boost of further confidence in the beliefs and thereby accelerates the process of reinforcing them . . . in a positive feedback loop of belief confirmation."[167]

A second, or additional, way in which beliefs are perpetuated is through socialization. Religious belief is the product of our evolution into magical thinkers and believers. However, as the next chapter makes clear, many sociologists and anthropologists argue that the human need to belong increases cultural support for magical thinking. E.O. Wilson argued along these lines that the evidence "points to organized religion as an expression of tribalism," first, because "every religion teaches its adherents that they are a special fellowship and that their creation story, moral precepts, and privilege from divine power are superior to those claimed in other religions." Second, "their charity and other acts of altruism are concentrated on their coreligionists; when extended to outsiders, it is usually to proselytize and thereby strengthen the tribe and its allies."[168]

5. Complex Socialization

Homo sapiens are an extremely social and tribal species. Most humans depend on their families in their formative years and even more on various social contacts in adulthood. Socialization has helped our species become the dominant one on the planet. Humans are not the largest or strongest animals; it was the ability to work together that facilitated their rise to the top of the food chain. Many factors facilitated this socialization, and it is difficult to say which are more important or, in some cases, which came first or promoted others, as is the case with hunting and language. The study of human socialization has focused on the physical evolution of the body, the social aspects of humans' closest living relatives (again, the great apes), and the theory of eusociality. The following discussion considers each of these topics in turn.

5.1 Physical Developments Facilitating Socialization

The act of walking upright, which is believed to have developed to facilitate mobility, perhaps on savannahs or areas with few trees, is among the factors that encouraged complex socialization. Walking upright not only frees the hands for other tasks but facilitates running for rapid escape from predators and, perhaps more importantly, rapid pursuit of prey. However, there was a penalty for greater mobility, specifically, anatomical change that included a smaller pelvis with a different position relative to the spine. The associated decrease in the size of the birth canal through which an infant's head must pass concurred with ongoing increases in the size of the brain and head. The evolutionary solution to this problem was for human young to be born at a point when their brains are less developed than those of most other mammals at birth. As a consequence, the rearing of a human child required much more time than is the case with other animals. By way of comparison, the gestation time required for a human brain to develop as fully as is the brains of most newborn mammals would be about 21 months.[169] The long period over which children are dependent on their parents is also a long period for the transfer of information. The trade-off here is that the parents must spend many years caring for vulnerable children, leaving them vulnerable themselves and, consequently, dependent on other humans for support and defense. Under such circumstances, the group is as important as the family for survival, which is to say that the family requires the support of a clan or tribe. In other words, survival then depends on the ability to socialize with those outside the family, fostering in individuals the need for a sense of belonging.

The large human brain naturally has an impact on social interactions in that it requires enormous amounts of energy and protein. This requirement was met by an increased diversity in diet to surpass the amount of protein available from sources such as beans or tubers.[170] One of the sources initially was carcasses, the remains of animals killed by other animals. Accessing this source of protein, however, meant living in the same environments as large predators and with the fact that their kills can be intermittent and unreliable. These drawbacks could be mitigated by killing animals rather than scavenging their dead bodies. One of the leading theories for the origins of language is in this transition from scavenging to active killing,[171] which requires not only sophisticated tools but extensive social organization.

5.1.1 Language & Belongingness

It is unclear which hominin species began to use language or when, but most experts date this development around 150,000 ya. As has been seen, language capability is concentrated in the parts of the brain known as Broca's and Wernicke's areas. These areas are not particularly pronounced anatomically and, consequently, do not leave distinguishing impressions in the endocasts of ancient skulls. Further, the development of language was probably not abrupt. Rather, language would have evolved from screeches, grunts, gestures, and exclamatory sounds that might be interpreted as expressions ranging from euphoric to disapproving similar to those made by the great apes. Some of these sounds would have evolved into basic nouns and verbs such as "go" and "stay."

With the demand for diverse sounds and types of communication, the range of sounds and inflections expanded as evolutionary pressure selected for changes in the larynx and the development of a second cavity, the pharynx (Figure 25). In the great apes and, probably, the earliest hominins, the larynx is directly behind the tongue and relatively small. The larynx of modern humans is not only larger but also much lower in the throat, and the pharynx further expands the range of tones that humans can produce.[172] Lacking these features, the great apes are limited to the sounds just described.

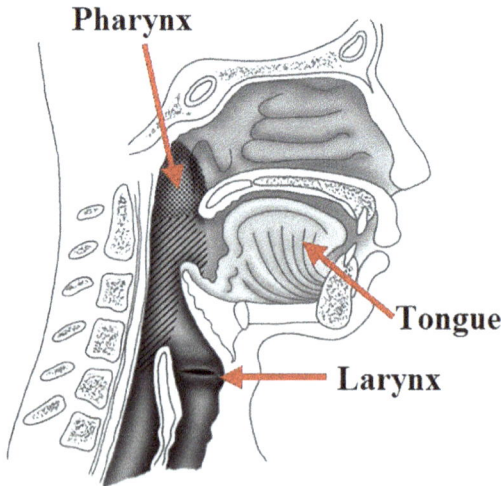

Figure 25. Anatomical structures associated with speech.
Image credit: drawing by Randy Mohr.

As noted, many anthropologists and neurologists have argued that language and tool-making developed in parallel or chronological proximity.[173] Dietrich Stout, for example, pointed, first, to the fact that both activities rely on and benefit from "general capacities," such as the collaboration of working memory and the Broca's and Wernicke's areas in communication through gestures. Thus, the expansion and coevolution of these capabilities would have supported both language and tool-using abilities. Second, Stout drew attention to "shared social context" involving "joint attention" and the contribution of language to "complex skill learning and cooperative activity" (i.e., hunting). His third argument in favor of a link between language and tools involves "shared neural substrates . . . including hierarchal combinations in the inferior frontal gyrus and action understanding in the cortical mirror-neuron system." Accordingly, tool use provided a "preadaptive foundation for specific aspects of the human language faculty."[174]

5.1.2 What Language Facilitates

Language, even when rudimentary, confers significant advantages, particularly in terms of coordinating the actions of multiple people. It is easier to organize a game drive, for instance, if you can tell the members of your group exactly where to go and wait for the prey and teach them to make traps. Leaders and their governing roles could be defined in greater detail. Trade would be easier with words (verbal symbols) for what a group wanted or had for trade. Without language, it is difficult if not impossible to construct anything of any complexity. Birds build nests without complex language, but nothing more lasting.

Picking up the discussion of the consequences of walking upright for childrearing, language also obviously facilitates teaching and learning various lessons, methods, and potential dangers over the many years that parents and children spend together. Parents can describe the past and acquaint children with their own parents and possibly grandparents. Children can learn that they are part of a family with a history, thereby reinforcing their sense of belonging and of their place in the world.

Language also opens up the future for discussion, whether tomorrow or the next season, and what must or could be done to ensure short- and long-term survival. It is unclear whether animals think about tomorrow, but, if they do, they certainly cannot share the details of their thoughts with other members of their group or make detailed plans. More advanced hominins developed the ability to use language and an enhanced imagination to discuss things, for instance, that do not but could exist, how or why a lightning strike occurred, or whether it might be possible to make a better hand axe.

Closely related to the future are other abstract concepts and even existential questions. With language, people could talk about dreams and their possible meanings. They could talk about death and what might happen after. They could talk about minds without bodies and spirits, making them the first spiritualists. Finally, language, as a form of symbology, would facilitate the creation and naming of eternal spirit figures.

5.1.3 Symbology and Art

By approximately 50,000 ya, members of the species *Homo sapiens* had become sufficiently equipped with language, autobiographical memories, self-awareness, a ToM, and empathy to express themselves symbolically in various forms of what we call art. Paintings began to appear on the walls of caves. The earliest known painting, in a cave on the island of Sulawesi in Indonesia, dates to about 43,900 ya.[175] Such caves were sometimes used as living spaces, but many were solely or primarily for the display of art. The anthropologist Barbara King is among those who have argued that these spaces may have been used for shamanistic, spiritual, or ritual purposes.[176] Most of the paintings are of animals, often in the context of hunting. It is unknown whether the animals were meant to depict previous or future hunts, to show reverence, or merely as art for art's sake. In any case, the fact that autobiographical memory allows for the use of past events in planning for the future seems relevant here. Indeed, Torrey suggested that these early hunting scenes are evidence of the kind of autobiographic memory that, according to him, would have provided *Homo sapiens* with a significant advantage over other hominid species.[177]

5.1.4 Hunting and Language

Hunting can be a solitary endeavor when the prey is small and the hunting ground is in an environment free of large predators. Solitary hunters in the Paleolithic through Neolithic eras, though, were at risk of becoming a big cat's lunch. Group hunting lowered this risk, but it required organization, which is to say, a cohesive social environment. The hunters needed to decide which direction to go and when, for example, in the morning or at night. They needed to decide when to attack and when to wait when they were setting up an ambush. A leader, a dominant man or woman, may have provided some direction. In any case, as discussed, communication was necessary for this kind of organization. Either hominins developed a language to facilitate hunting, or hunting techniques developed as a result of some rudimentary language. Either way, it seems apparent that language would have enhanced hunting in terms of deploying sophisticated techniques. Chimpanzees seem to hunt smaller apes in groups using known patterns and body language, but their techniques are nowhere near as sophisticated as those of humans.

Hunters in a group depend on each other for survival. As any combat veteran can attest, a bond quickly forms among the members of a group that is as strong as any family tie. The feeling of belonging that ties each member to the collective, to the point that one will die in defense of another is not seen in other species, at least not consistently. Further, it is this heightened sense of belonging that ultimately led *Homo sapiens* to seek the divine.[178]

5.2 The Social Behavior of Humans' Closest Living Genetic Relatives

The great apes, our closest living genetic relatives, offer some idea of the social organization of our earliest ancestors. Anthropologists who have lived with great apes have documented their social organization as well as their emotional capacity and other characteristics that hold the groups together. Thus, Barbara King identified five capabilities that she found to be essential to groups of gorillas related to the sense of and need for belonging and are also "fundamental building blocks of the religious imagination" or experience: empathy, meaning-making, following the rules, imagination, and consciousness.[179] The following discussion considers each of these capabilities in turn.

A. Empathy

Cognitive empathy, as King defines it, occurs when "one ape can put himself in another's shoes, intuiting the other's perspective or likely emotional state."[180] Primatologists have witnessed, for instance, aid given to an injured group member in the wild, reluctance to injure another gorilla in a lab setting, and support for a new mother of the same group in a zoo setting. In such cases, one gorilla perceives that another is in need.[181] This perception is a characteristic of a ToM, as mentioned, though far from sufficient to constitute one. The implication is that the possession of a ToM is not a binary phenomenon but rather, like other aspects of consciousness, manifests to a greater or lesser extent across species. The human capacity for empathy, which possibly carries over from earlier species, also forms part of the foundation of morality. That is, the understanding that it is unacceptable to harm another, except in defence, is the very basis for the Golden Rule cited earlier.[182]

B. Meaning-making

To cite King again, "Humans make meaning when we communicate."[183] Primates learn very early that, in order to belong, they must participate in certain social interactions involving communication that establish the individual's position within and expected contribution to the group. Belonging in this way provides a sense of meaning and purpose for the individual. In an evolutionary sense, the individual who experiences belongingness fits in well with the group, becomes a team member, and contributes, so the group selects and accepts those whom accepts the norms and rules of the group. Meaning-making requires two-way communication. In the great apes, communication primarily takes the form of arm movements and facial expressions. The limited vocalizations that apes make serve mainly to alert others to distress or danger. Humans establish meaning within groups by creating and following customs. Examples include special meals such as Thanksgiving and displays of belonging to select groups such as participation in parades on St. Patrick's Day and Independence Day. Events make meaning by enhancing, increasing, and solidifying feelings of belonging.

C. Following the Rules

Rules are conventions that groups create collectively. Great apes also establish and follow rules, many of which our species share. Thus, for instance, primate groups adhere to a "pecking order" governed by protocols such as assuming submissive posture when approaching a senior member of the group. Violating the rules can result in punishment. Similarly, for example, in the military, lower-ranking members must salute higher-ranking members first, and British "subjects" are expected to bow when approaching the nation's queen. In U.S. courts, the judge is addressed as "Your Honor." Such rules and protocols seem to be in force in all hominid groups. Members recognize that other members follow and respect the rules and customs and that they enhance the feeling of belonging.

D. Imagination

King defined imagination as "the ability to think beyond the here and now, to create a world in our heads that exists nowhere else."[184] Imagination greatly contributes to the feeling of belonging, for members can imagine the loneliness of not being part of the group. Like belonging, not belonging is part of the social construct that reinforces group cohesiveness. King cites several examples of great apes displaying imagination, including a chimpanzee raised by humans that pulled around an imaginary pull-toy and another in the wild that built a nest for an imaginary friend.[185] These observations demonstrated a basic capacity for imagination in apes but only to a limited extent beyond which language would be necessary.

E. Consciousness

For King, these four capabilities— empathy, meaning-making, following the rules, and imagination—are clear indicators of consciousness, all of which both apes and humans express. Again, without a shared language, it is impossible to determine whether another species has consciousness, at least in the way that humans do. Like intelligence and ToM, consciousness is not binary but rather an attribute or capability that stretches across a continuum, aspects of which appear in other species that are outside the scope of or entirely different from human understanding and imagination. In general, as stated earlier, any animal with a cortex may have some sort of consciousness.

5.3 Eusociality

E. O. Wilson proposed a theory for the evolution of species into complex societies that he called eusociality.[186] The theory delineates two steps that a species takes to become eusocial, which is to say, good at socializing. First, members of a group altruistically protect a long-term, defensible nest from enemies, whether predators, parasites, or competitors. Second, members of the group belong to more than one generation and share the labor in a way that sacrifices at least some of their personal interests for those of the group as a whole.[187] Further, at least some members of successive generations remain in the nest. This final step is attributable to the acquisition of an allele, that is, a modification of a single gene, that inhibits an innate urge to leave the group's nest to selfishly begin one's own.

Wilson based his theory on his study of insects such as ants, termites, and bees, but many of the characteristics of eusociality are applicable to *Homo sapiens.* Early in human evolution, the development of bipedalism, resulting in a longer gestation period, and the consequent need for group support described at the beginning of this chapter contributed to the selection for this allele. Further, at some point, group selection for increased altruism may have been the cause or result of an increase in the mirror neurons associated with empathy (for which see Section 4.3). In evolutionary terms, individuals who displayed empathy and altruism seem to have tended to reproduce more than those who did not over the generations. As alluded to earlier and discussed further presently, altruism is a potential source of conflict between the drive for self-preservation and the "greater good" of the group or tribe.

Early hominins, like chimpanzees, developed eusociality based on a life of hunting and gathering in groups that retained some of their offspring. Consequently, the major step towards eusociality was the establishment of permanent or semi-permanent "nests" or campsites that the group vigorously defended. The archaeological evidence for such sites occupied permanently by early hominins is slim and includes primarily the occupation of caves for long periods. However, the assumption that those who lived in these caves defended them vigorously and organized their communities around an extensive division of labor may be unwarranted. The first archaeological site that my review of the literature identified at which there is evidence for a division of labor was a permanent or semi-permanent "nest" at what is now Gobekli Tepe in southern Turkey dating some 11,500-9,700 ya, which I discuss further in the next chapter (see Figure 27). Still, it is only an assumption that the extensive construction and artistic carvings there would have required the combined efforts of construction workers, hunters, gatherers, and artisans—that is, a division of labor, and no sign of defensive structures have yet been unearthed at the site.

Although settlements with wooden defensive structures may have existed earlier, it seems to me that the earliest solid evidence yet discovered for permanent settlements characterized by a division of labor and defenses are sites such as Maidanetske in Ukraine. At this site, which I also discuss in the next chapter, there seems to be evidence for three walls (Figure 30) that may have been used for defensive purposes. If this is true, *Homo sapiens* may not have attained true eusociality until approximately 6,100 to 5,600 ya.

5.4 The Belonging Addiction, Eusociality, and the Limbic Paradox

The selection for belongingness as a contribution to the survival of groups has had biological impacts. The feeling of belonging is rewarded in and by the brain with the secretion of oxytocin by a nucleus within the reticular formation of the limbic system, which elevates the mood. This hormone, sometimes called the "love molecule," is produced, for instance, when people interact with babies and pets and when they engage in sexual intercourse.[188] The brains of all social species secrete oxytocin or a similar compound such as inoticin in bees, with the highest concentrations found in species that are sexually monogamous.

This biochemical reward facilitates the belongingness for not only members of the immediate family but the extended family and even individuals who are no blood relation. Humans can be affectionate, charitable, and inclusive. Thus, almost all cultures value acts of hospitality and helping strangers, and most religions promote the Golden Rule. Oxytocin seems to be at work in much of this behavior.

The belongingness reward system further facilitates and promotes altruism, which refers in this context to the sacrifice or compromise of the desires of the self for the benefit of the group or tribe. The result is constant internal conflict of morality or conscience as the members of groups ponder whether they should go fight to defend their group or focus on protecting those to whom they are related, or, less dramatically, whether they should report their income accurately and pay their fair share for public services or not report some of it in order to reduce their personal tax burden. According to E. O. Wilson, "Every normal person feels the pull of conscience, of heroism against cowardice, of truth against deception, of commitment against withdrawal. It is our fate to be tormented with large and small dilemmas."[189]

However, there is a tendency to overuse this adaptive survival feature, as with patternicity and cause-and-effect thinking. The members of our species want to belong not only to families and tribes but to most anything that does not seem threatening. We want to belong to large groups like towns and nations made up of people to whom we are not related. We want to belong to groups that seem exclusive and secret. We want to join Facebook and be "friended" and participate in groups that we've only just heard of or read about. We want to belong to and be part of nature. We want to belong to something bigger than ourselves. And, ultimately, we want to belong to "the divine," whatever we conceive it to be or are taught to conceive it to be. Belonging to the divine is the ultimate belonging because, in religions, it often solves seemingly unsolvable mysteries, such as concerns about death.

Humans internalize belongingness as an important part of self-identity and even self-worth. Self-identity and self-worth, in turn, being attached to belongingness, become attached to the limbic system at the basic emotional level. Belongingness fuels the drive to establish means of displaying group identity through language, clothing, or, most often, symbols. The members of groups create flags and monuments to symbolize their group identity and to which they can display their allegiance. The emotional attachment to belonging and the associated symbols triggers a defensive response in the event of a threat to the group or its symbols. Specifically, the limbic system responds to threats with the proverbial flight-or-fight response, releasing hormones that raise the blood pressure and stimulate the adrenal glands to produce adrenaline. Thus, the same limbic system that generates the oxytocin that encourages belonging to groups also facilitates the response to fear, including that felt when the member of a group is confronted with an "other" who is not a member. I call this situation the limbic paradox.

If *Homo sapiens* did not have checks on oxytocin-generating socialization, members of the species would be left vulnerable to infection, theft, and exploitation. Stephen Hinshaw is among those who have argued that mental modules evolved in the human brain to prevent this outcome.[190] Along similar lines, Richard Dawkins in *The Selfish Gene* advanced a theory of "memes," units of cultural transmission such as ideas, behaviors, attitudes, and values.[191] These "naturally selected" modules evolved to identify and marginalize individuals who look or act like the "other." The evolutionary pressure for the development of these modules was, specifically, the drive to avoid 1) signs of contagious disease or contamination, 2) people, cultures, and societies that appear economically or materially inferior and therefore covetous of the material possessions of other groups, and 3) those of different familial, tribal, or cultural status as defined by skin color, language, religion, and so on.[192] Such avoidance and repulsion results in dehumanization of the other. Once dehumanized, inhumane treatment becomes permissible. History shows the tendency of groups of humans to fear other groups and act on this fear by attempting to drive them away or even eradicate them. These exclusionary type memes historically may have been reinforced between 19,000 and 13,000 ya toward the end of the last ice age when climate was changing and there may have been contention for resources. Archaeological evidence in the Sudan indicates sustained minor warfare among hunter-gatherers.[193]

Stated more explicitly, the greatest problem with an exaggerated sense of belonging is exclusiveness. People tend to join groups that they believe to be superior to other groups and to avoid membership in groups that appear weak, immoral, or subservient to another group. There is a natural desire for recognition as one of the "best of the best." This exclusiveness can be particularly dangerous in religious contexts. Many religious groups believe that their members are the most moral or righteous among the broader population, with the implication that most of this population is immoral, unrighteous, or even evil, and, accordingly, deserving of avoidance or eradication. Belongingness gone haywire has been responsible for conflict, murder, war, and genocide.

5.5 Microevolution and Evo-Devo

Evolution and development, or "evo-devo," is an exciting subfield of genetics devoted to measuring the pace of genetic changes as individual members of a species develop and relating these changes to changes in a population.[194] During embrionic and later development cells produce proteins or long chains of amino acids corresponding to the DNA sequence of the cell. These proteins are necessary in the cell to facilitate skeletal, muscular, enzymes, and hormone construction. However, work has also identified gene proteins that are instrumental in coordinating cells as the bodies of animals take shape as well as so-called "master genes" and "toolkit genes." Recent research into this gene family has also identified the Pax family of DNA-binding proteins, which play a role in embryogenesis.[195] Among the DNA sequences involved is the microcephalin gene, one of the genes associated with brain size, which appears to have undergone significant change around 50,000 ya—a time when significant increases in elaborate burials and cave art appear in the archaeological record.[196]

Another of these toolkit genes associated with brain size, the ASPM gene, is believed to have undergone positive selective pressure about 7,000 ya. This is concurrent with when, the archaeological record shows the emergence of seasonal and permanent settlements such as Maidanetske, as discussed in the following chapter. These rapid changes in gene expression or modification or "accelerated evolution" can happen as rapidly as over the course of a generation.[197]

Another possible example of accelerated evolution in humans relates to the major lifestyle change from hunting and gathering to reliance on domesticated plants and animals that began around 10,000 ya. This change provided bearers of the new gene access to the large amount of protein found in animal milk. Specifically, the gene makes it possible to digest milk after breastfeeding, while those without it suffer from lactose intolerance.[198]

As has been seen (Section 5.1), the size of the brain had essentially reached a maximum owing to the size of the birth canal. However, internal changes remain possible and have occurred relatively recently. Hinshaw is among those who have argued that the changes in brain and cellular structure discussed in previous chapters may be the product of accelerated evolution, in particular, folds and fissures that increased the surface area of the cortex, larger parietal lobes and reduced occipital lobes to facilitate language, greater asymmetry between the halves of the brain related to specialization for handedness and language, and an increase in the concentration of high-speed von Economo neurons in the areas of the brain involved in decision-making.[199] These features are either not found in the great apes or are less developed.

5.6 Contemporary Hunter-Gatherer Social Behavior and the Ethereal

Ethnographic anthropologists have been studying contemporary hunter-gatherer tribes for over a century with respect to social organization, language, warfare, hunting techniques, and religion. Regarding religion, they have generally reported that hunter-gatherers lack organized religion with a centralized god or pantheon but practice instead shamanism, which involves engagement with a spirit world for the accomplishment of specific goals, such as healing the sick. Thus the shaman, or spiritual leader, claims to travel to a place where it is possible to interact with spirits during altered states of consciousness. These practices are less about faith and more about identifying the action felt necessary to accomplish a goal.[200] The shaman is not necessarily the chief or senior member of the group, but his or her powers often pass from father to son or mother to daughter.

The art produced by many hunter-gatherer societies depicts figures that combine the characteristics of animals and humans. These "theriomorphic" (after Greek θηρ, "beast, animal" and μορφη, "form") figures first appear in cave paintings dating to 43,900 ya in Indonesia (at Sulawesi) and 30,000 ya in Chauvet Cave in France. Some of the figures appear to be suspended in space, echoing claims by members of modern foraging groups that "movement through spirit worlds [is] often via flight rather than walking or running."[201] Many of the cave paintings are of animals that were contemporary with the artists, but some appear to depict imaginary combinations of animal features. It is unclear precisely what these figures represent, but they are certainly symbols meant to pass on information, trigger memories, or inspire feelings, just like language.[202] The individual painted images may not have had a spiritual or ritual purpose, but the set of relationships that the paintings suggest among actual animals, mythological animals, and theriomorphic figures are reminiscent of ceremonial environments, especially those associated with initiation into a group or subgroup.

The shamans in hunter-gather societies, then, exercise "jurisdiction over the treatment and diagnosis of a select set of problems, most frequently serving as a healer and diviner" of the group.[203] They fulfill their duties by interacting with what they believe to be spirits in a spirit world that they access in altered states of consciousness brought on by sleep or food deprivation or drugs, for shamans often have knowledge of psychotropic plants and fungi that affect the mind and, they believe, help them to heal members of their group or tribe. Their roles naturally change, however, when hunter-gathering communities start to settle down, as discussed in the next chapter.

6. Population, Agriculture, and Ritual

Hominins may have lived in wood or mud structures for a very long time but caves are the only residences that we have evidence for until approximately 21,000 ya. These *Homo sapiens* demonstrated the ability to build walls, such as the one in Theopetra Cave in Greece that appears to have been built to block the wind. In regions devoid of convenient caves, early humans mastered the construction of huts from mammoth bones, such as the one found at Mezhirich, Ukraine dating to about 18,000 ya.[204] Many of these structures seem to have been isolated, though some clusters of hearths have been found. Further, it is unclear whether these types of dwellings were occupied seasonally or year-round. If they built structures of wood or mud they have not survived or been discovered to date.

The end of the last major glacial period and subsequent warming (Figure 16) coincided with changes in socialization beyond those suggested by these finds. The frequency of cave paintings began to diminish as people began different forms of socialization. As human populations increased, so did competition for shelter as well as the resources acquired through hunting and gathering. Groups were faced with the choice of conflict or coexistence with competing groups, and there was natural selection for those willing to cooperate and who valued belonging. People began joining and living in groups much larger than the family-oriented hunter-gather bands in which they had previously tended to organize themselves. Those who secreted more oxytocin and displayed the behaviors of belongingness and acceptance would have adapted well to the increasing population and changes in living patterns. However, game animals and wild plants were insufficient to maintain large groups of people, so a new food source was needed.

6.1 The Farming and Spiritual Brain

Approximately 10,000 ya, at the end of the Late Glacial period, more drastic changes in the behavior of *Homo sapiens* began to appear. The most significant was the first agriculture. Further, clusters of dwellings separate from caves were being constructed. The precise reasons for this shift from a hunting-gathering to a semi-sedentary lifestyle may never be known, but the end of a prolonged glacial period seems to have been a significant factor.

In particular, this shift in the earth's climate brought the extinction of some of the species upon which *Homo sapiens* depended[205] as the warmer and wetter conditions, especially in the period from 18,500 to 13,000 ya, resulted in the replacement of open expanses of the vegetation that fed large mammals with forests and shrubbery. Consequently, more people were hunting than there were large fauna to feed them, so they looked for other sources of food, including the cultivation of plants and domestication of animals. The first farming was probably seasonal, and the first animal husbandry was probably nomadic. Populations began to settle around river banks, initially for the water available for drinking and irrigation, and later to transport goods to trade.

Sedentary living naturally differs from life as a hunter-gatherer. To begin with, the availability of resources to hunt and gather and the distance to reach them limits the size of a tribe or group of hunter-gatherers. An agricultural community, by contrast, enjoys safety in numbers from animal predators and external groups, and the size of the community is dependent on the efficiency of the farming techniques and quantity, quality, and diversity of the crops that it can produce. Each lifestyle fosters distinct protocols and rules for social engagement. After two million years of human social interactions and other aspects of life as hunter-gatherers, the shift in lifestyle by humans at this point could have selected for certain types of brain circuitry or configurations. The mindset of hunter-gatherers is less territorial since they follow migrating herds, remaining mobile and flexible. As just stated, the size of the tribe is limited to, or correlates with, the availability of game species and edible flora. The mindset of an agriculturist emphasizes territory and organization. First, a site for habitation must be chosen based on access to water for crops, animals, and people. After the crops are planted, they must be defended from wild grazing animals and other hominids and hominins, harvested, and, as far as possible, stored and preserved. It follows that some, if not all, members of the group need to stay near the crops for at least the duration of the crop cycle. Further, shelters must be built. Since the group is no longer dependent on wild game, its numbers can increase through reproduction and the incorporation of outsiders who share the group's goals. The individuals most adept at socialization would be the most successful in enhancing the group.

Torrey proposed that the dynamics and complexities of farming would have required additional changes to the brain. The most recent developments in the evolution of the human brain took place in the lateral prefrontal cortex and were probably supported by increases in the white matter tracts of the superior longitudinal fasciculus that provide connections to the parietal and temporal lobes, as shown in Figure 26.[206]

This theory is supported by the fact that the lateral prefrontal cortex and dorsolateral prefrontal cortex are the last areas of the brain to mature in children. That is, following the logic of "ontogeny recapitulates phylogeny," these structures must have developed most recently. Additionally, according to Torrey, the "main tasks of the lateral prefrontal cortex are planning, reasoning, problem-solving, and maintaining mental flexibility: these tasks are often referred to as the

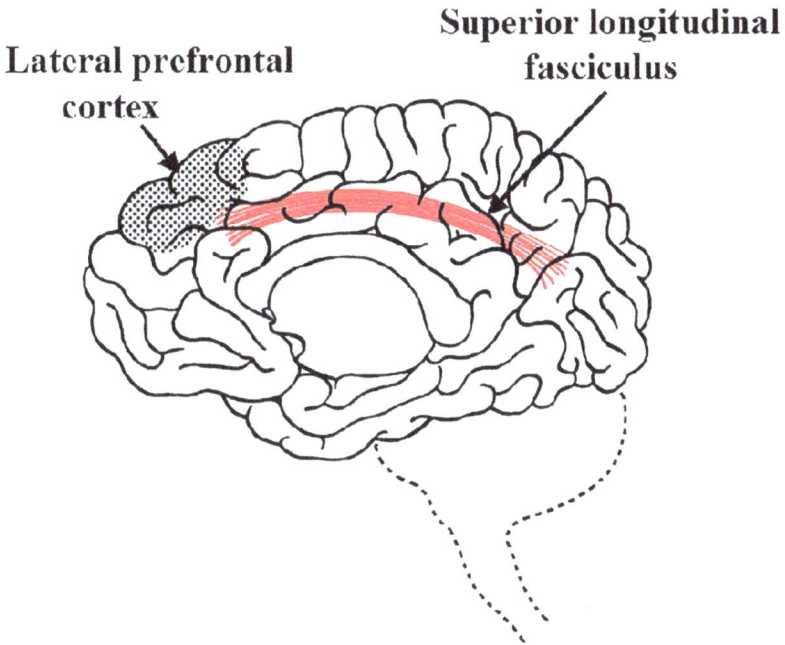

Figure 26. Relatively recent development in communication between parts of the brain.
Image credit: drawing by Randy Mohr.

executive functions of the brain."[207] These are also functions or tasks involved in farming.

As has been seen, the neurological foundations of ritual behavior, belief in disembodied minds, and spiritual life after death had already been laid by 11,000 ya.[208] These most recent developments in the lateral and dorsolateral prefrontal cortex supported by the high-speed association networks of the superior longitudinal fasciculus may have been the capstone for spirituality as well as facilitating the development of agriculture, again, around 7,000 ya (Section 5.5).[209] Seasonal collective activities began to take on specific social functions in specific spots. The purposes of these gatherings were various, with the archaeological evidence pointing to trade, rituals, and celebrations (or combinations thereof, e.g., festivities commemorating rites of passage). Mating opportunities among diverse groups may have been an added benefit.

The expression of spirituality may also have shifted significantly along with changes in art. Cave painting diminished around 15,000 ya, and large carved structures, requiring considerable team effort, began to appear around 11,000 ya.[210] An especially interesting site of this type is Gobekli Tepe in southeastern Turkey (Figure 27).

This site is of particular interest for the present discussion because it seems to have been carved from the bedrock and occupied during the transition from hunting and gathering to agriculture.[211] Since 60% of the animal bones found at the site were from gazelles and the remainder from vultures, cranes, ducks, and geese, these were hunter-gatherers. Nevertheless, the thousands of "grindstones, mortars, and carved stone vessels" found there too indicate large-scale food processing.[213] Hence, these people were either seasonal agriculturalists or somehow able to gather large quantities of wild grain. The work of carving the solid bedrock required an enormous amount of energy and free time indicative of a sufficient food supply and even a surplus. Since the majority if not all of their food was from hunting and gathering, a significant amount of the food was probably carried from some distance. Researchers have estimated that several

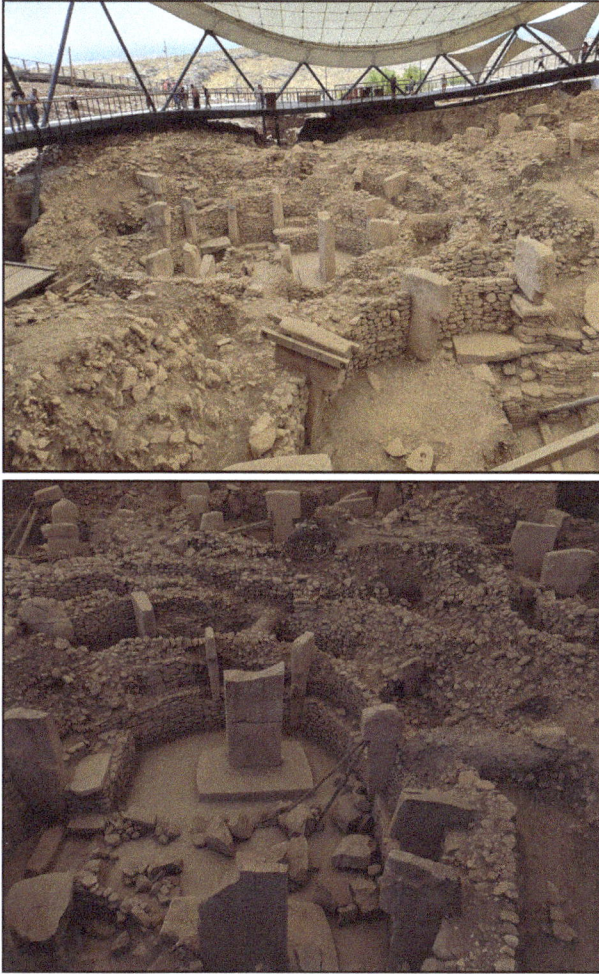

Figure 27. Gobekli Tepe in Turkey, 11,500-9,700 ya.
Image credit: photo by Arkadiusz Marciniak .
Image permission: Arkadiusz Marciniak

hundred people would have been required to carve and move the enormous stones at the site.²¹⁴ So, while the workers were carving the rock, many other people were hunting, butchering, and transporting food to the site, an arrangement that required both a dependable food supply and sophisticated organizational capabilities facilitated by complex language.

This complex was originally thought to serve primarily ceremonial purposes based on the extensive carvings and lack of household goods discovered. However, subsequent analysis indicates that the site served, probably from its founding, as a "home as well as church."[215] The T-shaped standing rock structures visible in the center of the photos in Figure 27 seem to have supported the roofs of dwellings as well as communal ceremonial spaces. This form of communal living would constitute a "nest" of the type that Wilson associated with the first step toward eusociality, that is, a social complex requiring a large investment of labor.[216]

Returning to an earlier thread of the discussion, for agriculture to succeed, permanent living quarters and collective activity are required, for crops need to be watered, harvested, stored, and protected from animals and other people.

The first agricultural communities may have not been much bigger than contemporary hunter-gatherer groups. However, prime agricultural land would be in high demand and represent a potential for conflict that would be most efficiently mitigated by sharing. Among the significant communal sites to develop in this early period of agriculture around 9,200-7,950 ya, the most renowned is Çatalhöyük in south-central Turkey (Figure 28). Faunal remains indicate that at least some of the residents of this village may have occupied the site throughout the year. Estimates of the population range from 3,500 to 8,000 over the 1,500 years that the site remained occupied.[217]

By 9,000 ya, people in some communities were keeping

Figure 28. Excavations at Çatalhöyük, Turkey, 9,200-7,950 ya.
Image credit: Photos by Arkadiusz. Marciniak.
Photos permission by Arkadiusz Marciniak

domesticated sheep and cattle and only periodically augmenting their diet with hunting.[218] The human brain had evolved sufficiently to plan seasonal animal husbandry and

Figure 29. Female icon from Çatalhöyük.
Image credit: photo by the author.

engage in other complex efforts to manage key resources. Notably, the excavations at Çatalhöyük have revealed distinctive burials, symbols, and icons, including a large carving of a well-endowed female figure seated in a chair flanked by two felines with a baby between its ankles (Figure 29). Anthropologists continue to debate the meaning of these symbols, but the icon seems to be the product of a stratified society, whether it represents a queen, deity, or the head of an extended matriarchal family, depicting someone or something that many if not all of the residents held in high esteem. At a minimum, this figure seems to exert an influence over dangerous animals as well as fertility. Although earlier portable female figurines have been found, this is one of the first that seems to be placed in a shrine installation associated with a permanent or semi-permanent settlement. Additionally, analyses of living spaces, communal spaces, burials, and other finds at Çatalhöyük indicate that, in the words of Kamilla Pawłowska, "people of the Middle phase were involved in elaborate rituals, symbolism, installations, and burials."[219] Accordingly, Torrey concluded that "by this time we were able to cultivate both plants and our spiritual selves."[220]

By 6,000 ya, communal sites the size of Çatalhöyük or larger were beginning to proliferate in what is now the northern part of Eastern Europe. Walled sites such as the aforementioned one at Maidanetske in Ukraine were becoming commonplace (Figure 30). The walls here, which formed

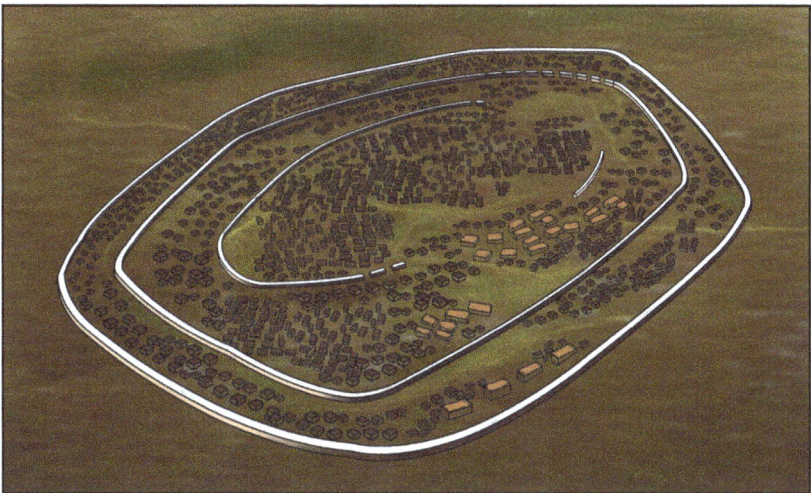

Figure 30. Reconstruction of excavations at Trypillia, a mega-site in Maidanetske, Ukraine, 6,100-5,600 ya.[221]
Image credit: drawing by Randy Mohr after site plan[222]

almost three concentric circles around the site, indicate the need for defense, whether against roaming bands, thieves, or well-organized competing megasites.

The Cucuteni-Trypillia culture inhabited what archaeologists call "megasites," which consisted of thousands of houses—3,000 in this case, though it is unclear whether they all existed at the same time or represent successive phases of construction, demolition, and rebuilding. Thus, the population of these communities, around 40 of which have been discovered so far,[223] tends to be hard to pin down, with estimates here ranging from 5,000-20,000 inhabitants.[224] Once more, it is unclear whether these sites were occupied year-round or seasonally. The people of Trypillia cultivated cereal grain and herded animals, apparently trading mainly in cattle, but they also hunted wild game. Notably, though, the evidence for hunting declined while that for the consumption of domesticated animals increased over time at many megasites.

Within another 2,000 years, Egyptian culture had appeared on the banks of the Nile River. We know a great deal more about this culture than any previous culture for two reasons. First, the Egyptians developed a written representation of their spoken language. Second, they were obsessed with the notion of life after death, so they built elaborate tombs both underground and within various pyramids, preserved corpses in the expectation that they would return to life in the future, and elaborated a mythology about the world beyond and how to live forever. Egyptians described this mythology in writing on the walls of dozens, if not hundreds, of tombs and in funerary texts commonly referred to in English as *The Book of the Dead* (Figure 31).

Figure 31. Egyptian pyramids.
Image credit: photos by the author.

Some ordinary Egyptians had their bodies preserved in tombs, while the pharaohs, believed to be, not only immortal, but gods, constructed enormous pyramids for the purpose. The Egyptians also built temples to honor these dead immortal rulers.

6.2 From Ancestor to God

As discussed in previous chapters, anthropologists and neurologists have suggested that dreams played a role in the development of both dualism and ancestor worship.[225] The existence of ancestor spirits seems very real to the members of many hunter-gatherer societies, and they employ the shamans discussed above to communicate with them. With the advent of farming, the ancestors of the people in agricultural communities or other spirits began to be conceived as possessing enormous power. J. Postgate argued that, from the very beginning, such powerful gods were associated with large populations.[226] If so, the critical mass of population necessary for an ancestor to become a god or for gods to take on far-reaching powers is unclear, but this development does seem to coincide with the shift to semi- or fully sedentary agricultural settlements. As alluded to earlier, many early cultures in the Middle East and Mediterranean region elaborated creation myths explaining the formation of the land on which they lived and their place in the cosmos. Often, these and associated myths involved multiple gods, each associated with a specific set of concerns that were paramount to their worshippers, such as rain, fertility, and defense. Again, these mythologies answered questions and solved mysteries.

The tendency toward anthropomorphization discussed above (Section 4.4) naturally extends to gods. Thus, the gods of most farming cultures were depicted as men and women endowed with supernatural powers and/or theriomorphic features (Section 5.6).[227] Further, irrespective of gender or form or associations, these gods possessed not only supernatural powers but also human emotions and intentions, indicating a kind of divine ToM. As Reza Aslan observed,

Whether we are aware of it or not, and regardless of whether we're believers or not, what the vast majority of us think about when we think about God is a divine version of ourselves: a human being but with superhuman powers.[228]

Discussion

My journey through our evolution, which began with a prehistoric hand axe, has been both challenging and exciting. I found that possession of a more resourceful, clever, and plastic brain than any other species is responsible for Homo sapiens' domination of the earth in a way that no other species has. This evolution was neither rapid, smooth, nor consistent, and many unanswered questions about it remain. The display of the capabilities needed to create a bifacial hand axe reflects a giant step in the development of the hominin brain. However, that giant step appears to have resulted, initially, in relative stagnation in terms of technological advancement. It is curious indeed that our ancestors continued to rely on this style of hand axe for 1.7 million years. There are indications that some degree of brain plasticity on several occasions led to the use of more diverse sets of tools and weaponry. Rick Potts and his team proposed that environmental challenges and resource scarcity fueled advances in tool use or the use of a greater variety of tools from roughly 500,000 to 320,000 ya in a specific part of the southern Kenyan Rift Valley. There is no evidence, however, that this spark of technological innovation spread to other areas. Rather, after this period, the hominins in the region reverted to use of the less-advanced collection of tools centered on the hand axe.

Perhaps most importantly, this journey has made me think about some concepts to which I had never thought about deeply enough before, concepts mentioned every day in familiar words such as "consciousness," "self," and "mind." I learned that these concepts are not binary—that is, either existing or not—but, rather, exist along a gradient or continuum from species to species. Virtually every mammal that has some cortex (and probably some non-mammals) experiences the personal, subjective sensation of information coming from both the external world and the subjective evaluation (emotion) of that information. My reflection on these deep concepts has changed my understanding of and attitude toward the other animals with which we share the planet. If all organisms with a cortex do, in fact, have dreams and memories, then we must share many subjective experiences with a great many other animals. On the other hand, the advent of a larger cortex and greater neural connectivity facilitated that which we call imagination which in turn allowed our ancestors to hold an image of a hand axe in its mind while napping two stones to create the actual tool. Or did persistently attempting to create a bifacial hand axe select for hominins with larger cortex? Either way, this imagination was the basis for an autobiographical memory that facilitated the creation of not only an individuals' history but a future and even an after life. This in turn has provided the basis for the concept that there is a mind that is separate from the body—dualism. It is this imagination that today allows us, and us alone, to conceive of an entire universe, or multiple universes, within our skulls.

Closely linked to complex tools such as the hand axe is the development of complex language. We may never know which came first and they may have even developed simultaneously. Given that they use adjacent neurologic motor circuitry they could be closely linked in development. Whether it was the advent of hunting or the discipline of teaching/learning napping techniques, the hand axe probably played a central role in the development.

I have drawn attention to the fact that evolution proceeds by fits and starts, producing changes that sometimes succeed and sometimes fail. Humans' exceptional ability to distinguish patterns from background noise and detect agents with potentially nefarious intentions has helped to ensure the species' survival in a competitive and dangerous world. Other animals have developed the capacity for patternicity and agenticity but, once again, not to the same extent. Patternicity together with our neurological reward system has endowed us with the need to ask and answer questions. Then, agenticity together with an enhanced imagination enabled us to conceive of an old man in the sky throwing down lightning bolts exemplified again by the Greek god Zeus and the Hebrew god YHWH.

In order to adjust to a congested world we also developed the neurological and hormonal capacity for belongingness, compassion, and love past the point of necessity to encompass concepts and activities that are both pleasurable (sports; throwing a hand axe in competition) and spiritual (religions; worshipping Zeus). That is, our successes may have brought with them some features that can be considered, in the lexicon of evolutionary biologists, spandrels or exaptations. These biological and sociological developments that now enhance fitness, rather than evolving through natural selection to play their current roles, allow for capabilities that, while no longer necessary for survival, have come to serve other useful functions.

I have also been intrigued and surprised by the fact that people relatively rarely make decisions based on the rational and objective collection and evaluation of data. Rather, our decisions are heavily influenced by our emotions, and even our collection of data is often subject to the biases established by our previous experiences. We tend to collect the data that support the emotions and biases that color our experiences when we first created a memory of the situation at hand or a similar one, also known as bias confirmation. Then, supported by data that are consistent with our emotional biases, we evaluate the options and make the decision that minimizes mental dissonance and maximizes neurological positive feedback and rewards. On the other hand, without the input of the emotional schema and limbic system, we are unable to make decisions. Given the enormous influence of drugs in the form of hormones from our limbic system upon our decisions and the evidence of deterioration of the mind and memory commensurate with known physical diseases such as Alzheimers and stroke, the support for the concept of dualism becomes less and less viable as research progresses. Although this realization can be unsettling to many, it emphasizes for all of us the need for the care, exercise, and continual development of one of our most important organs, our brain. These facts call into question the very possibility of free will. For example, with our patternicity and agenticity filters 'wide open,' and our need to have answers to mysteries, combined with confirmation bias, and the need for belonging we are vulnerable to joining conspiracy groups such as QAnon. Once convinced and emotionally committed, there are no alternative facts that can counter our limbic systems influence.

We can now also know that we are trapped between our limbic system's drive to love our neighbor and our limbic system's defensive guard (fear) to 'other' people who do not fit into our perceived tribe that enhances our feeling of belonging. As shortcuts to 'othering' we have created templates of color, beliefs, religions, sexual orientation, or social status in order to avoid, shun, and even destroy the 'other'. In many ways we have not progressed much since the advent of agriculture and settlement living. It seems that these othering memes may not be serving our species well in an increasingly crowded and globalized world where people of differing color, beliefs, religions, and sexual orientation need to live together.

On the other hand this research also revealed to me some

evolutionary advancements that could help to ensure the survival of the species. Our brains have evolved the plasticity to adjust to changing environmental and social conditions. The evolution of genetic mechanisms such as toolkit genes has endowed us with the inherent capacity to respond rapidly to changes in these conditions. The evolved ability to genetically adjust relatively rapidly is providing us with ever greater potential to modify our relationship with our environment and society. Additionally, with greater awareness and understanding of our emotional decision making process we may now proactively self-reflect on our own biases and motivations concerning both interpersonal relationships and our social context. The question may be whether we have the personal and collective will to make positive self-reflected decisions necessary for the improvement and survival of the species.

Appendix

Figure 32. Cheetahs.
Image credit: photo by the author.

Endnotes

1 The JVRP Survey is being carried out under the direction of Matthew J. Adams and Yotam Tepper (IAA License #: S-678/2016), who granted me permission to present the findings herein is courtesy of them.

The Acheulean type of hand-axe was named after the archaeological site near Saint-Acheul in northern France where the axe style was first found and has subsequently been found in Asia and Africa.

2 Curry (2018), p. 44.

3 Schlegel et al. (2013); Stout et al. (2015).

4 Stout et al. (2015), p.1.

5 Whiten et al. (2003); Bayern et al. (2018).

6 King (2007), pp. 173-175.

7 Schick (1999).

8 Thompson et al. (2019), p. 5.

9 Curry (2018), p. 43.

10 Thompson (2019), p. 2.

11 Torrey (2017), p. 48.

12 Ibid., pp. 27, 37.

13 Curry (2018), p. 43.

14 Torrey (2017), p. 9.

15 Ibid., pp. 14-16.

16 Cohen et al. (2016).

17 Norden (2007b), pp. 46-47.

18 Norden (2007a), pp. 69-70.

19 LeDoux (2019), pp. 257-258.

20 Ibid., p. 257.

21 Torrey (2017), pp. 49, 64. Mirror neurons were discovered in rhesus monkeys and are found in other primates to a lesser degree. As with many capabilities, the extent of the effect of these neurons is on a spectrum.

22 Nowell (2011), pp. 6-7.

23 LeDoux (2019), pp. 205-206.

24 Potts et al. (2020).

25 Ibid.

26 Torrey (2017), p. 69.

27 Darwin (1871), p. 319

28 Norden (2007b), pp. 198-199.

29 Ibid., pp. 200-201.

30 Ibid., p. 195.

31 Gazzaniga (2018), p. 215.

32 Torrey (2017), pp. 49-50.

33 Ibid., p. 49.

34 Ibid., p. 48.

35 Gazzaniga (2018), p. 220.

36 Ibid., pp. 219-220.

37 Hinshaw (2010), pp. 268-270.

38 Torrey (2017), p. 50.

39 Norden (2007a), p. 193.

40 Ibid., p. 194.

41 Ibid., pp.169-170.

42 I assume that his phenomena of data loss owing to the insufficient number of neurons occurs in other animals as well as humans and, consequently, that many animal species employ fill-in-the-blanks patternicity.

43 LeDoux (2019), p. 244.

44 LeDoux (2019), p. 244.

45 Ibid., p. 230.

46 Ibid., p. 231.

47 Torrey (2017), p.104.

48 LeDoux (2019), p. 296.

49 Ibid., p. 246.

50 Ibid., pp. 294-295.

51 Ibid., pp. 295, 300-301.

52 Norden (2007a), p. 193.

53 LeDoux (2019), p. 289.

54 Torrey (2017), p. 40.

55 Haxby et al. (2001).

56 Harrison & Tong (2009).

57 Horikawa et al. (2013).

58 Schlegel et al. (2013).

59 Ibid.

60 Stout (2015), pp. 9-10.

61 The controlling mechanism that initiates this transfer is not fully understood at this time.

62 I emphasize that, while references to dreams rarely appear in the same sentence as references to consciousness or cognition, it seems that many of the same brain areas defined as "working memory" are required for imagination and dreams.

63 Stout et al. (2015).

64 LeDoux (2019), p. 228.

65 Ibid., p. 229.

66 Norden (2007b), pp. 82-84.

67 Shermer (2011), p. 118.

68 Ibid., p.120.

69 Ibid., p. 5; this phenomenon is also called "pareidolia."

71 Ibid., p. 60.

72 Grossman (2021).

73 Blackmore & Moore (1994), pp. 91-103.

74 Marchlewska et al. (2017).

75 Skinner (1948).

76 Marchlewska et al. (2017).

77 LeDoux (2019), pp. 296-297.

78 Torrey (2017), p. 204.

79 Carroll (1965), pp. 181-182.

80 Benton et al. (2006).

81 E. Wilson (2012). p. 291.

82 Cantalupo & Hopkins (2001), p. 1.

83 Norden (2007b), pp. 28-29.

84 Wolpert (2006), p. 76.

85 Ashwell (2019), p. 192.

86 Norden (2007b), p. 20.

87 Kolodny & Edelman (2018), p. 5.

88 Ibid.

89 Klein & Blake (2002), p. 272.

90 Torrey (2017), pp. 78-79.

91 Gould & Lewontin (1979).

92 Gould & Vrba (1982).

93 LeDoux (2019), p. 239

94 Yong (2016); Pinker (1997), pp. 528, 524.

95 Kolodny & Edelman (2018).

96 Shermer (2011), p. 5.

97 Stanford (1999), pp. 10, 127, 160, 199-200.

98 Trumble et al., (2014).

99 Daw & Tobler (2014), p. 291.

100 Schulte-Hostedde et al. (2008).

101 Bernhardt et al. (1998).

102 Norden (2007b), p. 88.

103 Shermer (2011), p. 87.

104 Bourgoing (2001), p. 32.

105 Shermer (2011), p. 124.

106 Ibid., pp. 77-84.

107 Ibid., pp. 209-210.

108 Brotherton (2015), pp. 209-219.

109 Shermer (2011), pp. 259-261.

110 Torrey (2017), p. 72.

111 Eldredge & Biek (2019).

112 LeDoux (2019), p. 230.

113 Ibid., p. 231.

114 Ibid., pp. 257-258.

115 Ibid., pp. 256-257.

116 Torrey (2017), p. 49.

117 LeDoux (2019), pp. 306.

118 Ibid., p. 309.

119 Ibid., p. 303.

120 Torrey (2017), p. 84.

121 Hynes (2006).

122 Shermer (2011), p. 142.

123 Norden (2007b), pp. 35-36.

124 Ibid., p. 139.

125 Ibid., p. 128.

126 Norden (2007b), p. 139; LeDoux (2019), p. 351.

127 Norden (2007b), pp. 42-43

128 Damasio (1994), p. 70.

129 Descartes (1970), p.101.

130 Norden (2007b), p. 126.

131 Damasio (1994), pp. 35-36.

132 Ibid., p. 49.

133 LeDoux (2019), pp. 296-297.

134 Norden (2007a), p. 144.

135 LeDoux (2019), p. 354.

136 Shermer (2011), p. 124.

137 Jaffe (2019), p. 2.

138 Blair (2018).

139 Matsumoto et al. (2016).

140 Tokuyama et al. (2021).

141 Torrey (2017), p. 54.

142 Trinkaus & Villotte (2017).

143 Stringer & Gamble (1993), p. 94.

144 Jaffe (2019), p. 2.

145 Acharya & Shukla (2012).

146 Stickgold & Zadra (2020).

147 Gallagher & Firth (2003).

148 Liddell and Scott's Greek-English Lexicon, Simon Wallenberg Press, 2007, p. 63.

149 Genesis 8:20-21; New King James Version (1982).

150 Taylor, M (1999), p. 44.

151 Ibid., p. 32, 91.

152 "Religion: Gallup historical trends," (2021).

153 McNamara (2009) p. 203.

154 Franklin & Zyphur (2005), p. 1.

155 Winson (1993).

156 Torrey (2017), p. 118.

157 Franklin & Zyphur (2005), p. 1.

158 Kelemen (2004), p. 296.

159 Bulfinch (1970), p. 40. Psalms 18:13-14; New King

James Version (1982).

    160 Harris et al. (2008).

    161 Kapogiannis et al. (2009).

    162 Torrey (2017), pp. 62-64.

    163 Gazzaniga (2005), pp. 156-160.

    164 Kapogiannis et al. (2009).

    165 F. Wilson (2002).

    166 Lester (2002).

    167 Shermer (2011), p. 5.

    168 E. Wilson (2012), pp. 258-259.

    169 King (2017), p. 65.

    170 Antón,et al.. (2014), pp.2-3

    171 Thompson et al. (2019).

    172 Ghazanfarand & Rendall (2008).

    173 Stout et al. (2015).

    174 Nowell & Davidson (2011), pp. 160-162.

    175 Bower (2020), p. 9.

    176 King (2007), p. 142.

    177 Torrey (2017), p. 108.

    178 King (2007), pp. 50-56.

    179 Ibid., p. 56.

    180 Ibid., pp. 34.

    181 Ibid., pp. 33-38.

    182 Ibid., p. 40.

    183 Ibid. p. 44.

    184 Ibid., p. 58.

    185 Ibid, pp. 56-58.

    186 E. Wilson (2012), p. 110, defining eusociality as "true social condition." The prefix "eu" in ancient Greek is often translated as "well;" Greek "alethes" is usually translated as "truth."

187 E. Wilson (2012), pp. 140-141.

188 Norden (2007b), p. 60.

189 E. Wilson (2012), p. 290.

190 Hinshaw (2010), p. 318.

191 Dawkins (2006).

192 Hinshaw (2010), pp. 316-318.

193 Crevecoeur (2021), pp. 1-13

194 Ibid., p. 98.

195 Friedrich (2015).

196 Ibid., p. 99.

197 Hinshaw (2010), p. 100.

198 Ibid., p. 96.

199 Ibid., pp. 97-98.

200 King (2017), p. 138.

201 Bower (2020), p. 9.

202 King (2017), p. 136.

203 Singh (2018), p. 1.

204 Stringer & Gamble (1993), p. 189.

205 Lord et al. (2020), p. 3876.

206 Torrey (2017), p. 163.

207 Ibid., p. 162.

208 Ibid., p. 164.

209 Ibid., p. 164

210 Ibid., p. 103.

211 Curry (2021), pp. 25-31.

212 Curry (2008).

213 Curry (2021), p. 28.

214 Curry (2008).

215 Curry (2021), p. 28.

216 E. Wilson (2012) argued that ancient campsites would

have represented nests and the beginnings of eusociality. However, archaeologists have collected very little evidence concerning the length of time that early hominin campsites were occupied or the extent to which they were defended. Caves, as discussed at the beginning of the chapter, may have served as long-term nests.

217 Hodder (2020), p. 74.

218 Pawłowska (2020), p. 150.

219 Ibid., p. 151.

220 Torrey (2017), p. 164.

221 Hofmann et al. (2019).

222 Pappas, p.2

223 Gannon (2020).

224 "Mega-structures of Ukraine's Trypillia culture served as community centers," (2019).

225 McNamara (2009).

226 Postgate (1992), p. 112.

227 Less frequent were gods of animals or birds, such as the Egyptian falcon god Horace.

228 Aslan (2017), p. xiii.

Bibliography

Acharya, S. & Shukla, S. (2012). Mirror neurons: Enigma of the metaphysical modular brain. *Journal of Natural Science, Biology, and Medicine,* 3(2), 118–124.

Alcock, J. E. (2018). *Belief.* Prometheus.

Antón, Susan C., Potts, Richard, Aiello, Leslie C. (2014) Evolution of earlyHomo: An integrated biological perspective. Science, 345 (6192). 1236828pp. doi:10.1126/science.1236828

Ashwell, K. (2019). *The brain book* (2nd ed.). Firefly Books.

Aslan, R. (2017). *God: A human history.* Random House.

Bernhardt, P. C., Dabbs, J. M., Fielden, J. A., & Kutter, C. D. (1998). Testosterone changes during vicarious experiences of winning and losing among fans at sporting events. *Physiology of Behavior,* 65(1), 59–62.

Benton, T. R., Ross, D. F., Bradshaw, E., Thomas, W. N., & Bradshaw, G. S. (2006). Eyewitness memory is still not common sense: Comparing jurors, judges and law enforcement to eyewitness experts. *Applied Cognitive Psychology,* 20(1), 115–129.

Blackmore, S., & Moore, R. (1994). Seeing things: Visual recognition and belief in the paranormal. *European Journal of Parapsychology,* 10, 91–103.

Blair, R. J. R. (2018). Neurobiological basis of psychopathy. *British Journal of Psychiatry,* 182(1), 5–7.

Bower, B. (2020, January 18). Oldest known figurative art discovered. *Nature.* doi: 10.1038/s41586-019-1806-y

Bulfinch, T. (1970). *Bulfinch's mythology,* Thomas Y. Crowell Company.

Brotherton, R. (2015). *Suspicious minds.* Bloomsbury Publishing.

Bourgoing, J. de. (2001). *The Calendar.* Harry N. Abrams, Inc., Publishers

Cantalupo, C., & Hopkins, W. (2001). Asymmetrical Broca's area in great apes. *Nature,* 414(6863), 505.

Carroll, L. (1965). *Alice's adventures in Wonderland and Through the looking glass.* Airmont.

Cohen, A. O., Breiner, K., Steinberg, L., Bonnie, R. J., Scott, E. S., Taylor-Thompson, K. A., Rudolf, M. D., Chein, J., Richeson, J. A., Heller, A. S., Silverman, M. R., Dellarco, D. V., Fair, D. A., Galvan, A., & Casey, B. J. (2016). When is an adolescent an adult? Assessing cognitive control in emotional and non-emotional contexts. *Psychological* Science, 27(4), 549–562.

1. Crevecoeur I., Marie-Hélène Dias-Meirinho, Antoine Zazzo, Daniel Antoine, François Bon. New insights on interpersonal violence in the Late Pleistocene based on the Nile valley cemetery of Jebel Sahaba. *Scientific Reports*, 2021; 11 (1) DOI: 10.1038/s41598-021-89386-y

Curry, A. (2018). Imaging the past. *Archaeology*, 71(2), 42–45.

Curry, A. (2021). Last stand of the hunter-gatherers? *Archaeology, VOLUME 74, Number* 3,(May/June 2021), p. 24–31.

Curry, A. (2008, November). Gobekli Tepe: The world's first temple? *Smithsonian Magazine,* Nov. 2008 VOLUME(ISSUE), 000–000. https://www.smithsonianmag.com/history/gobekli-tepe-the-worlds-first-temple-83613665/

Damasio, A. (1994). *Descartes' error.* Penguin Books.

Darwin, C. (1871). The descent of man and selection in relation to sex. In Hutchins, INITIAL., Adler, INITIAL., & Brockway, INITIAL. (Eds.) *Great books of the Western world* (p.319). PUBLISHER.The Modern Library

Daw, D. D., & Tobler, P. N. (2014). *Value learning through reinforcement: The basics of dopamine and reinforcement learning: Neuroeconomics.* Elsevier.

Dawkins, R. (2006). *The selfish gene.* Oxford University Press.

Descartes, R. (1970). *The philosophical works of Descartes* (E. S. Haldane & G. R. T. Ross, Eds.). Cambridge University Press.

Eldredge, S., & Biek, B. (2019, September). Glad you asked: Ice ages—what are they and what causes them? *Utah Geological Survey*, PAGE NUMBERS, URL, OR DOI https://geology.utah.gov/map-pub/survey-notes/glad-you-asked/ice-ages-what-are-they-and-what-causes-them/

El Mehdi Sehasseh, *et al.* Early Middle Stone Age personal ornaments from Bizmoune Cave, Essaouira, Morocco; *Science*

Advances Vol. 7, No. 39

Franklin, M. S., & Zyphur, M. J. (2005, January 1). The role of dreams in the evolution of the human mind. *Evolutionary Psychology*, 1–25.

Friedrich, M. (2015). Evo-devo gene toolkit update: At least seven Pax transcription factor subfamilies in the last common ancestor of bilateral animals. *Evolution & Development*, 17(5), 255–257.

Gallagher, H. L., & Firth, C. D. (2003). Functional imaging of "theory of mind." *Trends in Cognitive Sciences*, 7(2), 77–83,

Gazzaniga, M. S. (2005). *The ethical brain.* Dana Press.

Gazzaniga, M. S. (2018). *The consciousness instinct.* Farrar, Straus and Giroux.

Ghazanfarand, A. A., & Rendall, D. (2008). Evolution of human vocal production. *Current Biology*, 18(11), R457–R460.

Gould, S. J., & Lewontin, R. C. (1979). The spandrels of San Marco and the Panglossian paradigm: A critique of the adaptationist programme. *Proceedings of the Royal Society* B: Biological Sciences, 205(1161), 581–598.

Gould, S. J., & Vrba, E. S. (1982). Exaptation—A missing term in the science of form. *Paleobiology*, 8(1), 4–15.

Grossman, L. (2021, July 3 & July 17). Galactic arc may challenge cosmology. *Science News*, p. 9 PAGE NUMBERS, URL, OR DOI

Harris, S., Sheth, S. A., & Cohen, M. S. (2007, February 27). Functional neuroimaging of belief, disbelief, and uncertainty. *Annals of Neurology*, 63(2), 141–147.

Harrison, S. A., & Tong, F. (2009). Decoding reveals the

contents of visual working memory in early visual areas. *Nature*, 458(7238), 632–635.

Haxby, J. V., et al. (2001). Distributed and overlapping representations of faces and objects in the ventral temporal cortex. *Science*, 293(5539), 2425–2430.

Hinshaw, S. P. (2010). *Origins of the human mind*. The Great Courses.

Hodder, I. (2020). Twenty-five years of research at Çatalhöyük. *Near Eastern Archaeology*, 83(2), 72–79.

Hofmann, R., Müller, J., Shatilo, L., Videiko, M., Ohlrau, R., Rud, V, Burdo, N., Dal Corso, M., Dreibrodt, S., & Kirleis, W. (2019). Governing Tripolye: Integrative architecture in Tripolye settlements. *PLoS ONE*, 14 (9), e0222243. doi: 10.1371/journal.pone.0222243

Horikawa, T., Tamaki, M., Miyawaki, Y., & Kamitani, Y. (2013). Neural decoding of visual imagery during sleep. *Science*, 340(6132), 639–642.

Jaffe, A. (2019, July 17). A look in the mirror neuron: Empathy and addiction. *Psychology Today*. https://www.psychologytoday.com/us/blog/all-about-addiction/201907/look-in-the-mirror-neuron-empathy-and-addiction

Kapogiannis, D., Barbey, A. K., Su, M., Zamboni, G., Krueger, K., & Grafman, J. (2009). Cognitive and neural foundations of religious belief. *PNAS*. doi: https://doi.org/10.1073/pnas.0811717106

Kelemen, D. (2004). Are children "intuitive theists"? *Psychological Science*, 15(5), 295–301.

King, B. (2017). *Evolving god*. University of Chicago Press.

Klein, R. G., & Edgar, B. (2002). *The dawn of human culture: A bold new theory on what sparked the "Big Bang" of human consciousness.* Wiley.

Kolodny, O., & Edelman, S. (2018). The evolution of the capacity for language: The ecological context and adaptive value of a process of cognitive hijacking. *Transactions of the Royal Society.* doi: https://doi.org/10.1098/rstb.2017.0052

LeDoux, J. (2019). *The deep history of ourselves.* Viking.

Lester, T. (2002, February). Oh gods. *The Atlantic,* pp. 37–45.

Liddell and Scott's Greek-English *Lexicon,,*Simon Wallenberg Press, 2007, p63

Lord, E., Dussex, N., Kierczak, M., Vartanyan, S., Gotherstrom, A., & Daken, L. (2020). Pre-extinction demographic stability and genomic signatures of adaptation in the woolly rhinoceros. *Current Biology*, 30, 3871–3879.

Marchlewska, M., Cichocka, A., & Kossowska, M. (2017). Addicted to answers: Need for cognitive closure and the endorsement of conspiracy beliefs. *European Journal of Social Psychology*, 48, 109-117.

Matsumoto, T., Itoh, N., & Nakamura, M. (2016). An observation of a severely disabled infant chimpanzee in the wild and her interactions with her mother. *Primates*, 57, 3–7.

McNamara, P. (2009). *The neuroscience of religious experience.* Cambridge University Press.

"Mega-structures of Ukraine's Trypillia culture served as community centers." (2019, September 26). *SciNews.* http://www.sci-news.com/archaeology/mega-structures-ukraines-trypillia-culture-community-centers-07630.html

New King James Version. (1982). Thomas Nelson Bibles.

Norden, J. (2007a). *Understanding the brain: Lectures* 1-18. The Great Courses.

Norden, J. (2007b). *Understanding the brain: Lectures* 19-36. The Great Courses.

Nowell, A., & Davidson, I. (2011). *Stone tools and the evolution of human cognition.* University Press of Colorado.

Pappas, S. (2019, September) Mysterious Megastructures of the Elusive Tripolye Culture Unearthed in Ukraine. *LiveScience.* https://www.livescience.com/mysterious-megastructures-tripolye-culture.html

Pawłowska, K. (2020, September). Towards the end of the Çatalhöyűk East Settlement: A faunal approach. *Near Eastern Archaeology*, 83(3), 146–154.

Pinker, S. (1997). *How the mind works.* Penguin.

Postgate, J. N. (1992). *Early Mesopotamia: Society and economy at the dawn of history.* Routledge.

Potts, R., et al. (2020, October). Increased ecological resource variability during a critical transition in hominin evolution. *Science Advances*, Potts et al., Sci. Adv. 2020; 6 : eabc8975 21 October 2020 PAGE NUMBERS, URL, OR DOI

Religion: Gallup Historical Trends. (2021). *Gallup.* https://news.gallup.com/poll/1690/religion.aspx

Schlegel, A., Kohler, P. J., Fogelson, S. V., Alexander, P., Konuthula, D. & Tse, P. U. (2013). Network structure and dynamics of the mental workspace. *PNAS.* doi:10.1073/pnas.1311149110

Schulte-Hostedde, A. I., Eys, M. A., & Johnson, K. (2008). Female mate choice is influenced by male sport participation.

Evolutionary Psychology, 6(1), 113–124.

Shermer, M. (2011). *The believing brain*. St. Martin's Press.

Singh, M. (2018, April). The cultural evolution of shamanism. *Behavioral and Brain Sciences*, 1–62.

Skinner, B. F. (1948). Superstition in the pigeon. *Journal of Experimental Psychology*, 38(2), 168–172.

Stanford, C. B. (1999). *The hunting apes*. Princeton University Press.

Stickgold, R., & Zadra, A. (2020, December). The biological function of dreams. *The Scientist*. https://www.the-scientist.com/reading-frames/opinion-the-biological-function-of-dreams-68184

Stringer, C., & Gamble, C. (1993). *In search of the Neanderthals: Solving the puzzle of human origins*. Thames and Hudson.

Stout, D., Hecht, E., Kheisheh, N., Bradley, B., & Chminade, T. (2015, April). Cognitive demands of lower Paleolithic toolmaking. *PLOS* One. doi: https://doi.org/10.1371/journal.pone.0121804

Taylor, M. (1999). *Imaginary companions and the children who create them*. Oxford University Press.

Thompson, J. C., Carvalho, S., Marean, C. W., & Alemseged, Z. (2019). Origins of the human predatory pattern. *Current Anthropology*, 60(1), 1–23.

Tokuyama, N. T., K. Poiret, M. L., Iyokango, B., Bakaa, B., & Ishizuka, S. (2021, March). Two wild female bonobos adopted infants from a different social group at Wamba. *Scientific Reports*, 11, 4967.

Torrey, E. F. (2017). *Evolving brains, emerging gods.*

Columbia University Press.

Trinkaus, E., & Villotte, S. (2017). External auditory exostoses and hearing loss in the Shanidar 1 Neandertal Supporting Information Appendices. *PLOS* One. Dataset.

Trumble, B. C., Smith, E. A., O'Conner, K. A., Kaplan, H. S., & Gurven, M. D. (2014). Successful hunting increases testosterone and cortisol in a subsistence population. *Proceedings of Biological Science*, 281(1776), 20132876.

von Bayern, A. M. P., Danel, S., Auersperg, A. M. I., Mioduszewska, B., & Kacelnik, A. (2018). Compound tool construction by New Caledonian crows. *Scientific Reports*, 8, 15676.

Whiten A., Horner, V., & Marshal-Pescini, S. (2003). Cultural panthropology. *Evolutionary Anthropology*, 12, 92–105.

Wilson, E. O. (2012). *The social conquest of Earth.* Liveright Publishing.

Wilson. F. S. (2002). *Darwin's cathedral: Evolution, religion, and the nature of society.* University of Chicago Press.

Winson, J. (1993). The biology and function of rapid eye movement sleep. *Current Opinions in Neurobiology*, 3, 243–248.

Wolpert, L. (2006). *Six impossible things before breakfast.* W. W. Norton & Company.

Yong, E. (2016, January). We're the only animals with chins, and no one knows why. *The Atlantic.* https://www.theatlantic.com/science/archive/2016/01/were-the-only-animals-with-chins-and-no-one-knows-why/431625/

Glossary of Terms

Acheulean tradition: The Acheulean type of hand axe was named after the archaeological site near Saint-Acheul in northern France where the style was first discovered and has subsequently been found in Asia and Africa.

agenticity: the concept that every event, including the movements of objects, must have a cause.

apocalypticism: a belief system that anticipates a future imminent event, such as the arrival of a savior figure, the end of the earth, or even the overturning of an election.

Australopithecus: a genus of the family *Hominidae* of bipedal primates with both ape-like and human features that lived in Africa during the Pliocene and Pleistocene (approximately 4 to 1 mya).

basal ganglia circuits: a collection of subcortical nuclei involved in motor control.

dorsolateral prefrontal cortex (DLPFC): a functional (as opposed to anatomical) area of the brain involved in the executive functions of coordinating activities and functions that include working memory, planning, reasoning, and inhibition.

dualism: the perception that the mind is separate from the body.

endocasts: representations of the interior spaces of skulls.

eusociality: literally, "true social condition." Members of a eusocial group belong to multiple generations and divide labor in an altruistic manner.

evo-devo: *evolution* and *devel*opment that involve relatively rapid change.

exaptation: a term describing features that now enhance fitness but were not shaped by natural selection to serve their current roles.

functional magnetic resonance imaging (fMRI). A technique using strong magnetic fields to detect energy or activity in targeted areas of the brain based on changes in blood oxygenation level-dependent (BOLD) data.

granular cells: very small neurons that have various functions depending on their locations in the brain. Granular cells are the most numerous of the various types of neurons.

Homo erectus: literally, "[walking] upright person," a species of the genus *Homo* ("person" in Latin) that appeared about 2 mya and is believed to have diverged into new species about 500,000 ya.

Homo habilis: literally, "handy person," a species of the genus that appeared about 2.1 mya. and originally thought to have been the first to use tools. Subsequent research, however, has associated the discovery of tools with earlier *Australopithecus.*

Homo sapiens: literally, "wise person," anatomically modern humans, who originated 200,000-300,000 ya in Africa and spread beyond the continent about 70,000 years ago.

lateral geniculate nucleus (LGN): an area in the thalamus in the brain that serves as a mapping platform and relays signals traveling from the optic nerve to the occipital lobe.

left superior frontal gyrus (LSFG): an area in the front of the brain involved in self-awareness.

lunate sulcus: a crescent-shaped depression touching on the occipital, parietal, and temporal lobes of the brain.

monism: the concept that the human mind is connected to neurological processes and that there is only one substance.

neocortical tissue: a term, incorporating the Greek νεο ("new"), *cortical- Latin (bark)* referring to the outer layer of brain tissue.

Oldowan tools: a set of artifacts named after the region in eastern Africa where were they were first identified.

"ontogeny recapitulates phylogeny": the notion that the development of an embryo proceeds through stages that reflect the evolution of the species.

Paleo-anthropologist: a term, derived from Greek παλαι ("old," "ancient") and ανθρωπος (man"), referring to an expert in the early development of the genus *Homo.*

paleocortical tissue: a term, also incorporating Greek παλαι *(old or ancient) and cortical- Latin (bark),* referring to the outer layer of brain tissue.

Paleo-neurologist: a term, also incorporating Greek From the Greek παλαι (old or ancient) and νευρον ("sinew," "tendon"), referring to an expert who studies the evolution of the nervous system.

paleontologist: *From the Greek παλαι (old or ancient) and* λογιοσ (skilled words or study of words), referring to an expert who studies past geologic periods from the perspective of fossils.

pareidolia: the capacity to perceive specific images in a random or ambiguous visual pattern.

patternicity: the tendency to find meaningful patterns in both meaningful and meaningless data.

plasticity: the ability of the brain to change, modify, or, essentially, re-wire itself.

posterior parietal cortex (PPC): an area of the brain involved in planning movement, spatial judgment and reasoning, and directed attention.

pottery sherds: pieces of broken pottery.

precuneus (PCU): an area of the brain involved in such complex functions as environmental perception, recalling episodic memories, and formulating strategies involving imagery.

retinal ganglion cells (RGCs): the long neurons that make up the optic nerve. RGCs process the light entering the eye and transmit the processed signal to the brain.

retinotopic map: a representation of the visual data received in the brain's visual centers from the neurons of the optic nerve.

schema: a pattern of thought involving the organization of groups of data and their relationships to each other. Examples include an individual's preconceived ideas and "world view."

shaman: an individual considered to possess the ability to access spirits for the primary purpose of solving a problem (e.g., curing a sickness).

superior longitudinal fasciculus: A fibrous tract of neurons in the brain that connects the frontal, occipital, parietal, and temporal lobes.

teleology: an explanation for a natural phenomenon in terms of a design or purpose.

thalamus: an area in the center of the brain with many nuclei that relays sensory data to the cerebral cortex.

von Economo neurons (VENs): cells, also known as spindle neurons, found primarily in the anterior cingulate and anterior insular

Acknowledgments

From Hand Axe to Zeus to QAnon is an ongoing project in that all of the sciences relevant to human evolution continue to evolve with respect to the understanding of the development of our species. The project started as an effort to satisfy my curiosity and expanded as I gradually realized the scope, volume of research, and knowledge that was already available on the relationship between stone tools and the development of the brain. I found myself particularly intrigued by the continuing migration of concepts from the realm of philosophy and psychology into the lexicon of neurology. Further, the development of fMRI technology for visualizing the brains of humans as well as animals as they perform tasks, including tasks that involve the imagination, has resulted in exponential growth in our understanding of the mind.

The project has not been a solo endeavor, but has benefited from the gentle contributions of folks who are very important to me. My fiancée, Louise Garrett, a Licensed Clinical Social Worker, has provided not only valuable feedback but also critical advice about remaining as objective as possible when discussing sensitive topics. The artist and archaeologist Randy Mohr has been my patient and critical colleague, creating many of the drawings. Another archaeologist and friend, Dr. Bob Susick, offered early editorial feedback and has continued to encourage me to complete the project through to publication. My longtime close friend Larry Geluso reviewed the penultimate draft exhaustively, supplying numerous comments and thoughts and pointing out places for clarification, all of which enhanced the final product. My close friends and colleagues Dr. George H. Goldsmith, Scott Roberts, and Brian O'Connor also provided valuable early comments and support. Also, my close friend Catherine Tucker provided critical 'proofing' comments. Of course, the editors were absolutely critical in the production of the final document. Finally, Dr. Jose Gaudier, a practicing neurologist, helped with important comments and technical verification of evolving concepts.

About the Author

Jim Bucko is an anthropologist and writer with an energetic curiosity into several related disciplines including neurology, psychology, geography, geology, cosmology, and theology. Currently retired, his primary activities include field archaeology, research, writing, and making presentations to interested groups.

His formal training includes a Bachelor of Arts from Pennsylvania State University with majors in Anthropology and Psychology, and a Master of Interdisciplinary Studies in Archaeology from George Mason University. He is also a member of the Registry of Professional Archaeologists (RPA), the American Society of Overseas Research (ASOR), the Society of Biblical Literature (SBL), and the American Humanist Association (AHA).

Jim has twenty-five seasons of experience in archaeological field work in Italy, Turkey, Israel, and Jordan with an emphasis in Roman era sites and general field surveys.

He has also traveled extensively on all seven continents and to over 100 countries. An avid hiker and climber who enjoys expedition climbing in Africa and South America, he has summited Aconcagua, Chimborazo, and Kilimanjaro among others.

A lifelong researcher and adventurer, Jim is addicted to his curiosity and the rewards of learning.